SONGS OF LOVE AND WAR

SONGS OF LOVE AND WAR

The Dark Heart of Bird Behaviour

Dominic Couzens

B L O O M S B U R Y
LONDON • OXFORD • NEW YORK • NEW DELHI • SYDNEY

To my sister, Cathy Hawkes.

Bloomsbury Natural History
An imprint of Bloomsbury Publishing Plc

50 Bedford Square
London
WC1B 3DP
UK

1385 Broadway
New York
NY 10018
USA

www.bloomsbury.com

First published 2017

British Library Cataloguing-in-Publication Data
A catalogue record for this book is available from the British Library.

Library of Congress Cataloguing-in-Publication data has been applied for.

ISBN (hardback) 978-1-4729-0991-6
ISBN (ePUB) 978-1-4729-0992-3
ISBN (ePDF) 978-1-4729-1588-7

2 4 6 8 10 9 7 5 3 1

Chapter illustrations by Marianne Taylor

Typeset in Bembo Std by Deanta Global Publishing Services, Chennai, India
Printed and bound in Great Britain by CPI Group (UK) Ltd, Croydon CR0 4YY

Contents

Prologue

We used to call them adventures. Now that my son is eleven, we call them hikes. Either way, we like to be outdoors, away from people, in the hinterland where wildness and safety dwell together. We were dropped off beside the trunk road that cuts the New Forest in half, and we set off to cross wood and mire, heath and scrub to reach the village of Sway 15km (9.94 miles) to the south. The forest was damp and breezy, quintessentially March, the spring still largely in the mind. There were no brilliant colours yet, only fresh ones. The roadside daffodils in Burley shone but didn't catch fire. Instead, the landscape was full of subtleties: the watery greenish-yellow of hazel catkins, the militarily upright spikes of dog's mercury, the smooth, sunshine-dappled bark of beech – green against green, brown against brown. It might seem impossible to overlook a tree that is pink against purple, but alder is that tree, its dense lattices of buds, catkins and bark keeping their colours secret.

Hiking is about curiosity, to find out what lies around the next bend, or at the edge of the wood. It can, of course, be about companionship too, and for us today, father and son, it was two generations and two minds coming upon the same scenes together and reflecting in our different ways. Samuel had not noticed the charcoal-like tips at the end of ash twigs before. On the other hand, I had long forgotten about the attractions of deep mud, something to be ploughed through rather than edged around. For a while, I borrowed his gift of childhood and revelled in the simple joy of getting stuck. Mud and water provide ways to convert a gentle walk into something resembling an adventure.

This being March, I was surprised at one point to see a swallow fluttering determinedly across the side wind, hacking northwards. It was an early arrival, even for the New Forest, in the deep south of England, and the swallow seemed to be struggling in the cold, insect-less air. If swallows could ever regret things, which we are pretty certain that they don't, then this one might have been berating itself about its impetuosity in getting here so soon. Swallows usually fly with effortless abandon, their wings rowing the air currents like expert oarsmen so that, every few moments, they can topple one way or another, zooming up or down, banking, sweeping, playing. Today this swallow's light touch was a hindrance. It kept low and its head down.

It was hard not to compare its journey with ours. We were on a leisurely 16-km (10-mile) stroll, for our own pleasure, in each other's company, knowing that there was a train to catch at the end to take us back to a warm home. As for the swallow, it had almost certainly flown across the English Channel this morning, closing in on its destination, perhaps now only a day or two away. Its marathon northward journey, at least 12,000 km (7,456 miles) from its beginnings in the Cape region of South Africa, had probably begun in late January. What must it have seen on its travels? Maybe the swallow saw elephants and giraffes, a mass of humanity, the great rainforests of the Congo Basin. Along the way it probably snacked on insects that have not yet been described by science, and perhaps never will be. Its shadow would have passed over trees and clearings never seen by the human eye. We can never know how it registered such things. We cannot say it was interested or excited, and we cannot say if it wasn't, because we don't know how swallows think. We can only guess as to its own level of curiosity, because we cannot measure curiosity in a swallow. We can guess at its motivation, though. The swallow was on a mission, to reach its breeding grounds in good time. It was on its own, unaided, doubtless well ahead of its rivals. Probably that is all that mattered.

The swallow's lot seemed stern and burdensome, especially compared to the joy of our benign mission. We could stop any time we were tired, if we wanted to, and there were shops for food well within reach. We didn't have to reach our destination at all; we had only chosen it yesterday. Yet for a swallow, not reaching its goal would be disastrous. If it fell short of last year's territory (or its birthplace, if it was migrating for the first time), through tiredness or hunger or even disorientation, it would be starting off the breeding season in a new place, not knowing the ropes and having never met the neighbours. Such things matter for a bird and can make the difference between it reproducing or not. The swallow's stakes were as high as ours were minimal.

And as we contentedly traipsed through the early spring forest, carefree and curious, I kept being reminded of the burdens and challenges of the birds crossing our path. A pair of crows flapped past, calling to each other. If they hadn't been inky-black and flying with slow, stealthy wing-beats you could have dismissed them as early season lovebirds, cawing with friskiness. And indeed, crows are usually faithful and attentive partners. Yet theirs is an introverted world, where togetherness is a buffer against a hostile neighbourhood. There is mutual aggression from magpies, casual violence, and that's not all. Some of a crow's worst enemies are other crows. Settled pairs are often harassed by individuals without a territory, and may even be raided by a flock. Sometimes their eggs and young are destroyed. They must always be on their guard, always ready to defend their territory, primed to fight. They must be watchful and remain healthy. A single bad day might alter the course of their life.

One wonders what goes through crows' minds as they follow the course of normal existence on a day like today, a dose of Englishness with blustery air and pretend cold. Do they live on the edge? Do they perceive the constant threat of disaster? Or do they simply react to things? Their lives are shorter and more brutal than ours, but it seems hard to believe that a crow could be riddled with anxiety. It is far easier to imagine that, even when relaxed, they can react with great speed to changing situations.

I said the weather was unremarkable that day, but that didn't mean it was agreeable. As we edged towards Wilverley Inclosure on an old, disused railway line the first drops of rain began to tickle our faces, and we could see drizzle against the conifers on the top of the hill. This was not showy or torrential, just highly professional, moistening rain, lightly spraying the puddles and making the birch buds drip. It didn't stop the birds singing; indeed, the great tits were uttering repeated 'tee-cher' songs with their usual gusto. Great tits provide the most authentic sound of early spring, ringing in the new season in January, just after Christmas, but in the context of that rainy day in March, their cheeriness sounded forced and false, like holidaymakers making loud jokes under grey skies at the seaside.

For all its status as everyday rain, from which experienced hikers are generally immune, precipitation itself is not necessarily a good thing for birds. While the male great tits chimed, their mates would be feeling the stress of the wet and wind and relative cold. Householders who feed birds tend to sympathise most with their visitors in the depths of winter, but the early spring can be a dangerously in-between time. It is the ecological equivalent of a black hole in the finances, when all capital is spent and investments haven't yet borne fruit. Almost all the countryside's production of seeds, fruits and berries takes place in the autumn, with little real replenishment, so stocks of plant produce, upon which many birds depend, dwindle through the winter and into the spring. On the other hand, the new bloom of biological production, in the form of insects and other invertebrates, hasn't really started. The months of February and March, therefore, can be challenging, even for the healthiest and most resourceful birds.

So March is a time of vulnerability for a number of resident birds, and the vagaries of weather can have a startling effect. Inclement conditions might not actually kill birds, but can certainly hamper what they are trying to do. Females, in particular, need to use the pre-breeding season to get into condition; soon they will be producing eggs, but they won't if they cannot feed well. Male birds need to defend territories, but if they are underfed they won't be able to compete. A short spell of poor weather and the cold, in particular, will reduce the number of insects that are active, making them harder to find. Wind in the trees makes it harder to look for insects. That was obvious as we watched a great tit trying to forage. Every so often it could do nothing except cling on to the yellowish-budded twigs. Eventually, it gave up and flew off.

Things were still more turbulent up in the treetops. As we stepped into Wilverley Inclosure, each of the tips of the spruces and pines lurched to and fro in the breeze, dancing a flamenco with some invisible whirling partner. What must it be like in the upper branches on a blustery day for a bird? If you were ever a fan of *Star Trek*, you will remember how the cast had to run back and forth across the set every so often to simulate some kind of intergalactic storm. I had this image in my mind's eye as I spotted a coal tit flitting in the upper storey. These tiny birds feed on the minutest of invertebrates among the conifer needles, requiring precision to insert their bills into the right places. They also need to hop from perch to perch accurately. Each weighs 9g (0.3oz) or so, less than the weight of three pennies. What, I wondered, would happen if a coal tit lost its grip? Out of control, it would be whisked away violently by the wind – to what kind of fate? Yet somehow this individual managed to cling on, using its modified, fissured feet for a strong grip, presumably working in the shelter behind the patches of needles. The contrast between our earthbound, muddy progress and the coal tit's treetop acrobatics could hardly have been more profound, yet we were only separated in space by a few metres.

We live a parallel life to our birds. We are neighbours in space, and when birds visit our gardens we feel as though we have a window onto their world. Yet really we have no idea of how they live, and little concept of what their motivations really are. We cannot appreciate their difficulties, nor the brutalities of their lives. We don't live close to them at all.

A small finch, a siskin, flew over us, giving its typical sighing and rather weedy call. It was followed by several others. Samuel and I heard it and saw it, but whether the siskin saw us or heard us, I could not tell you; nor does it matter, except that, as birdwatchers and bird-appreciators, we might sometimes subscribe to a false vanity. The siskin was a wild bird living in the New Forest, commuting from the embracing, feather-like midriffs of tall spruces to make visits to streamside alders where it would feed on seeds and drop down to the water to drink. The point is that it and its kind live their lives devoid of any intended attention to humanity. The New Forest might possibly be large enough to host birds that have never even seen a human – certainly, this must happen in Scotland where siskins are common and is probably the norm in Norway or Sweden. Siskins, along with other birds, are immune to humanity. They might flee from us, but not in any different way to their retreat from other perceived threats. In the garden, our proximity to birds and their habit of taking food that we set out makes us think that we might matter to them. But we don't. Wild birds are indifferent to us; they are immune to our charms. It is a conceit to think that our lives might intertwine with those of the birds in the garden. Nothing in humanity would be affecting the diary of our siskin. If it wasn't a local breeding bird, it would be setting off to Scotland in a few days' time – or perhaps heading across the North Sea – regardless of any interactions with us. Birds get on with their lives, and their choices are not our choices.

The more birds we saw on our happy walk, father and son together, the more I came to realise how proximity is no path to understanding. Most of us are vaguely aware of just how removed we are to the wild things around us, but few of us are acutely sensitive to how great is the divide between ourselves and wild creatures. The physical divide is clear enough: birds are more closely related to reptiles than they are to us – indeed, they are the true descendants of dinosaurs. Most are small, and two of their limbs have evolved for flying; their senses and reactions are, generally, sharper. However, the physical differences are easier to appreciate than the distinctions in our lifestyles.

Take the concept of a 'home', for example. Human beings, even most of the poorest, have a home, which gives them a roof over their head and space, however meagre, that they can call their own and return to every night. A home, though, is a much looser concept for a bird, and often quite an alien one. The children's books that charmingly suggest that a nest is a bird's home are typically wide of the mark, at least for British birds. Take the blue tits, for example, that brightened up the early spring branches with vivid cobalt and yellow, like living Christmas decorations. Pairs of blue tits only occupy their breeding hole from April until June, at the most. The 'family home' is a dangerous and vulnerable place, the noise and activity and the comings and goings of the parents also carrying the risk of attracting unwanted attention. After breeding, the family splits up forever, the young dispersing and probably never seeing their parents again. Siblings may join a flock together in the late summer, but their association won't last long; some will disperse further, most will die. If they do survive, the siblings will join the local pool of blue tits, all of which are in competition for resources. The adults join the same pool and, although the same male and female might well meet up again in the spring and attempt to breed together, they won't associate in the meantime and certainly won't share anything. They will have separate roost sites, as will all the young, wherever they find themselves. And although each bird might well return to the same sleeping place for many consecutive nights, every bird will need alternatives in case of trouble. None are entirely safe and secure. During the daytime blue tits don't use their roosting holes; they have no living rooms or places exclusively theirs. The wild is invariably a shared space.

Compare this to a human family. However wet and cold Sam and I might get on our New Forest stroll, we are fortunate to have a place of warmth and comfort, and family togetherness, to which we can return at any time of day. Our family unit is, God-willing, a stable and peaceful foundation from which we hope that our offspring (I also have a daughter, Emmie) will thrive. If we have to move house we tend to do this by choice and at leisure, something that would be a luxury for almost any bird.

Another aspect of human life, at least here in the developed world, is how we are able to enjoy a degree of good health and a life expectancy that is rare indeed in raw nature. Obviously, we are subject to the vagaries of disease and mishap, but the degree to which this has reduced, compared to human society only fifty years ago, is astonishing. Once again this experience is at odds with all the wildlife we see close at hand, including birds. It is a marvellous thing and a rich blessing, but it divorces us from the reality of life in the wild, and on our doorstep.

By human standards, individual birds face grim statistics and their successes seem to be against the odds. In fact, on our jaunty New Forest walk, full of hope and sprung footsteps, I could not bring myself to recall the numbers and add clouds on to our benevolent horizon. However, the fortunes of birds are brutal. Take the life-expectancy estimates of some of the birds that we passed by: starling, two and a half years; blackbird, a fraction less than that; blue tit, one and a half years; and robin, pathetically, just a year and a month. The statistics for young birds are very much less, suffice to say that of a brood of ten blue tits, only one can expect to survive to breed the following year. One in about 10,000 blue tits would live to be my son's age.

These statistics are shocking. It means that, on any given birdwatching trip to a place such as the New Forest, especially in the autumn (when there is a higher proportion of young birds around), only a tiny minority of those individuals could be seen again on the same trip to the same place the following year. Of all the blue tits that hatch, 37 per cent die as fledglings. Death stalks birds in many guises, but there are three true grim reapers: starvation, predation and disease.

There is little more comfort to be had in the interrelationships between birds. In human society, especially in the affluent west, we expect support from the body of society and from the state. Starvation is rare, disease and ill health are kept in check, and violent crime – roughly equivalent to predation – has a place on the margins of our reality. In the society of birds, though, there are few safeguards. Birds do associate in flocks for mutual benefit, but competition is an altogether more dangerous foe than it is in human society. In the world of songbirds, society's losers lose everything – the chance to feed, the chance to reproduce, the chance to survive.

We will explore many of these things in the chapters that follow, but the overall message of this book is that birds are never safe and their lives are brief and violent. Bird behaviour has a dark heart. Any bond between us and our wild bird 'neighbours' – even the most familiar songbirds that we see in our gardens each and every day – is a figment of our imagination. Most of us enjoy watching birds and glimpsing into their lives, but the feeling is not mutual.

Back in the New Forest, somewhere near Holmsley, Sam and I got lost. The woods all looked the same, as did the bogs and the clouds. The map could have been a hand scrawled impression of a treasure island, for all the help it was to us. We trawled through mud and turf and came to a narrow but deep river, its water the colour of fudge. The weather had given up the pretence of benevolence and it was now raining constantly, giving an evening feel to mid-afternoon. The map confusingly declared the existence of a footbridge, but that could have been somewhere else entirely. To go south we needed to cross the muddy-banked river.

A little further down an oak had fallen in middle age, by our good fortune directly over the river. We looked at each other and, without needing to say anything, we climbed onto its slippery hulk. With great care we crawled across, glancing down at the less than inviting waters, our senses enlivened by the unexpected intrusion of minor risk. We crossed and jumped down, still lost but cheered by the necessity to be slightly brave. After half an hour we spied a woman approaching, accompanied, or rather led by a gangling, ungainly dog that loped all over the path. We broke the masculine taboo of asking directions. The woman declared that she had walked the circuit for years, but then confessed that she neither knew the name of the Inclosure we were in nor knew how to get out of it. She reminded me of the sort of character who appears in an adventure novel to make a land sound magical and strange, like the tardy White Rabbit in *Alice's Adventures in Wonderland*.

Then suddenly the map sprung into sense, like a computer rebooting, and we reoriented to the south-east, sure of every sunken step. The forest hissed with the sound of driving rain and mist softened every edge of every leafless tree; the countryside became blurred as if the rain was causing a freshly painted canvas to run. Sway couldn't arrive soon enough.

As we approached the edge of the village there was a commotion. Songbirds called urgently, giving their generic, stressed 'see-see' high-pitched warning of aerial danger, and small bodies scattered. A sparrowhawk, all flexed wings and muscle, sped past in a shallow dive, eager for a feathered meal. We didn't see exactly what happened but there might have been a death. If not there was certainly a near miss, a moment when a bird's reactions saved it at the last millisecond. You might call it a drama, or even an adventure. It also happened to be stark, real life. Sam and I witnessed it, a sentence in our notebook of experience, soon forgotten. We could not, even in our imagination, really know how such an incident played in the minds of perpetrator and potential target. It was nature inured to us, ourselves inured to nature. We were joyful and satisfied with our walk, as one of nature's dark shadows passed by.

We as people will never know how a bird actually perceives its own life, its own brevity and struggle. We can, though, understand more than we do about how birds are different from us. This book is about the reality of a bird's life, as illustrated by a selection of behaviours, from singing to fighting and from breeding to migrating. It is about how the ticking heart of a bird's life is unusual and surprising but, above all, darker than we might think.

A note on the birds covered in this book

This book is mainly about songbirds, the diminutive neighbours that are most familiar to us – thrushes, starlings, tits, finches and so on. That is partly because songs and singing are a major theme of the book, rather than quacking or squawking. It also keeps the subject matter contained to a set of birds rather than all British species, which would be too extensive to do justice to here. It is about the lives of birds, but only parts of their lives and only certain species.

The term 'songbird' can be used technically, referring to a collection of species with a well-developed sound-producing mechanism in the windpipe (the syrinx) as well as a specific arrangement of toes on the foot (three pointing forward and one back, or 'anisodactyl'), allowing them to perch efficiently. They are members of a single order of classification known as Passeriformes, and the members are often referred to as Passerines. This group includes almost all small landbirds – those mentioned above, plus robins, wrens, warblers, flycatchers, sparrows, swallows, wagtails, larks and so on – as well as some larger ones such as crows.

On occasion, I have strayed into covering the lives of some species that aren't Passerines. Once or twice this is because of personal experiences, but more often it is because certain species are both very familiar and pleasantly noisy, the most obvious examples being the cuckoo and the pigeons/doves. For the purposes of this book, they are honorary songbirds.

CHAPTER ONE

Awakenings

It's before dawn in a picture-postcard Dorset village in the exhilarating month of May, and one of nature's wonders is about to unfold. It should be a time of awe and anticipation. It should be an opportunity to be imbued with Keats-like lyrical wonderment. But it's strange how often you have to suffer a little to experience the best of the outdoors, and today it is bitterly cold, with a strong wind, and the dark is unfriendly and unpromising. You can imagine on a day like this the dawn itself being reluctant to appear, let alone the dawn chorus starting up. The sun will eventually rise on what seems like an ungrateful morning.

I have come here to commentate on the dawn chorus for the villagers of Cerne Abbas and find the participants huddled on a street corner like emperor penguins withstanding an Antarctic gale – and probably furnished with as many protective layers. I feel an uncomfortable surge of responsibility, knowing that these people are only here because they expect to witness something amazing. But will the birds deliver? Will they decide at the last moment to stay in their roosts? I can almost imagine a mocking article in the local newspaper – *Birds Keep Quiet in Dawn Chorus No-show.* This could be ugly, as well as freezing.

A wit breaks the silence. 'The birds will need to sing to cheer themselves up,' she muses, and there is a ripple of merriment.

Salvation is quickly at hand, though. Without warning a few notes from a blackbird carry above the wind a few rooftops away. They sound more like a gentle suggestion than a song, under the breath and exploratory, but they are an opening salvo nonetheless. It is barely past 4am, but we must now hurry to the edge of the village, to the bridge over the stream. The dawn chorus is beginning.

Unlike many wildlife spectaculars – wildebeest massing on African plains, lions hunting, even red deer rutting in Scottish glens – the dawn chorus is a backyard miracle. In the UK, for example, throughout much of the spring, from late January to June, only a warm bed and inertia separates it from its potential human audience. It is possible to hear the dawn chorusalmost anywhere, from the heart of a city to the depths of a forest, and this proximity perhaps reduces the urge in us to experience it to the full, just as Londoners are less likely to visit the Tower of London than a tourist visiting the city for a few days. Another consequence of proximity is that people misunderstand the dawn chorus. It doesn't even have the correct name because it is really a pre-dawn chorus, which plays out in the darkness before the sun rises.

Here in Dorset, though, we have beaten the urge to stay in bed and are ready. The cold has quickly cleared our heads and the darkness tunes up our ears to full alertness. We might easily have missed the robin starting if it hadn't. The song of the robin is everything the bird is not – gentle, wistful, inoffensive. The singer is some distance away; this always seems to happen with dawn choruses, the birds close at hand never seem to start off first. Perhaps they sense our presence. One study has shown that birds remain quiet when they know there is a predator in the vicinity.

The overture begins: robin, blackbird, robin again, blackbird again; nothing else for the moment. Robin and blackbird share a characteristic that explains why they often begin 'chorusing' earlier than the rest of the songbirds. Both are adapted to feeding on the ground in the deep shade, where it is often dark. Consequently, they have eyes that, in proportion to their body size, are larger than those of many other species, and can presumably better detect the first infinitesimal increase in pre-dawn light intensity. They are natural early risers.

As it happens, robins and blackbirds have similar songs. They sing a phrase, pause, and then sing a different phrase to the last one, with much variety. The difference is in the pitch, with the robin sounding shrill and tinny and the blackbird fluty and tuneful. You can imagine both as interviewees, answering questions that you cannot hear during their pauses. And in a way, all dawn-chorus birds are answering questions, asked of them by a real audience. But more of that later.

The dawn chorus has begun with a lone voice and a whisper, but the first voice is like the first drop of rain in a shower. When a blackbird starts, its neighbours are compelled to start also; within a few minutes, there is a blackbird on a rooftop, another in the woods up the hill and yet another close to the bridge, all singing, without hesitation. The robin, too, is quickly answered by a close neighbour. For a while, each robin fills its neighbour's silences until a third joins in. Around and about there is a jumble of robins, a jumble of blackbirds.

It is time for the percussion, and in this case, it comes from a carrion crow. Crows are not necessarily early risers, so perhaps our presence has disturbed this one. There is no question of a whisper from this irrepressible character. It is technically a songbird in terms of its biological affinities, and it is capable of uttering an impressive range of quiet, intimate sounds. But not now. Instead it welcomes the day with an irritable tirade of three caws, belted out, a mouthful of swearing from the treetops. Never mind the wind and the cold and the dark, though – carrion crows always bawl and nobody ever seems welcome, at any time of day.

The shout from the canopy shatters the overture and seems to hasten other highly strung singers into action. From the undergrowth comes a voice that is almost always unseen, disembodied, even during the daylight. The wren would be a stranger to humankind were it not for its loud voice. The song is bold, fast and energetic but, in contrast to those of blackbird or robin, it consists of the same phrase repeated again and again, with only minor variations. This makes the wren the first of the day's 'slogan singers', with the same beginning, middle and end. The wren seems to blurt out its song and it never seems quite to keep control of its own voice. I love to compare its speed and zest to an excited commentator describing the last few steps in a 100m sprint. Having said that, it would be the same, or very similar commentary every time.

Wrens are never the earliest in a dawn chorus, at least not in my experience. You might expect them to be, because they, like robins and blackbirds, are birds of shade and cover, used to operating in the dark. They live in the tangled alleyways of thick vegetation close to the ground – so much so that their scientific name, *Troglodytes*, means cave-dweller. You might expect them to have similarly large eyes in comparison to other species, as do robins and blackbirds. But it seems they don't. So at the cusp of dawn wrens are not up with the robin or the thrushes, but in the roll-call of a dawn chorus they always lag behind. In contrast to robins and thrushes, wrens roost in holes, so perhaps they do not perceive the change in light intensity and instead take their cue from other singers?

Of course larks – and – especially our common and widespread species the skylark – are famous for being early risers. Although today there is no lark singing at Cerne Abbas – it could be too windy for them – I have personal experience to back this up. When experiencing the dawn chorus on the nearby Dorset coast, the skylark is always the first songbird to be heard. It seldom seems to bother about daybreak, seemingly happy to sing in the dark to a background of fading stars. Of course, the skylark usually performs in the air – I say usually because it is perfectly normal, especially in inclement conditions, to hear one on the ground – so perhaps it spots the light advancing on the eastern horizon. Then again, some cue must send it aloft in the first place.

The skylark's song is proverbial. It has been celebrated through literature and music, always with delight. John Clare (1793–1864) said of the skylark that:

Up from their hurry, see, the skylark flies,
And o'er her half-formed nest, with happy wings
Winnows the air, till in the cloud she sings,
Then hangs a dust-spot in the sunny skies,
And drops, and drops, till in her nest she lies,

And Shelley (1792–1822) goes all guns blazing into a kind of ecstasy:

Hail to thee, blithe Spirit!
Bird thou never wert,
That from Heaven, or near it,
Pourest thy full heart
In profuse strains of unpremeditated art.

More recently, at a time when perhaps the real bird is heard less often than it used to be, the skylark is commemorated in the piece by Ralph Vaughan Williams, 'The Lark Ascending'.

This piece consistently tops the poll in the radio station Classic FM's annual Hall of Fame voted for by listeners each Easter. What is perhaps surprising, in view of the overall reputation of the composer, is that piece is top of a pile of 300 classical pieces beating Mozart, Beethoven, Tchaikovsky and the rest. The piece does represent the skylark well, although you would never actually learn the bird's song from listening to it, yet you can somehow feel the clear skies, the freedom and the purity. *The Lark Ascending* doesn't come close to the power and melodic quality of many of its rivals, but it wins because people are nostalgic for summer skies filled with skylark song, the vocal blossom of the spring and summer. They also love it, perhaps, because the piece reflects a certain Englishness.

What is most unusual about the skylark's song in a biological sense is its continual nature. Most bird songs are sung in what you might call 'sentences', which don't last long. The skylark's effort, by contrast, is like a tap that turns on when the bird first rises from the ground and only turns off when the singer is a couple of metres from landing again. In the meantime, it will have slowly gained height to perhaps 30m (100ft), poured out its song while oscillating up and down slightly in the air, and perhaps circled above its territory. The whole performance can last up to five minutes, although three is average, and for all that time the outpouring never ceases for a moment. The lark's song is ecstatic and fits the mood of a hot summer day perfectly – certainly not an icy morning in Cerne Abbas.

Once they have recovered from its emotional impact, the question everybody always asks about skylark song is 'How do they sing for so long without running out of breath?' And indeed, estimates of the amount of air in the skylark's respiratory system suggest that they should indeed come to a shuddering, breathless halt. The reason that they don't is that they take extraordinarily fast 'mini-breaths' between notes, or between suitable groups of notes. These intakes of breath may last for a mere twelve to fifty milliseconds, which is why humans cannot hear them. There isn't time for the bird to replenish oxygen in that time, only to inhale again what it just exhaled, but the mini breaths do at least allow the bird to finish what it is singing. When singing a long trill, where the notes repeat more than about thirty times a second, the bird has to shift to proper inhalation. But once again, the so-called 'inter-trill breaks' are very short to our ears.

Hearing skylarks and thinking of continuous outpourings of song, I am left wondering whether some people must have the capacity to take mini breaths, such is the outpouring from their vocal cords. Sometimes I wonder, after putting the phone down to a loquacious friend, whether there was ever an inter-trill break.

My musings on skylark song are broken by a new voice in the swelling chorus, that of the chaffinch. Where the skylark – and the blackbird – are virtuosos well beyond grade eight in their metaphorical music exams, the chaffinch is stuck in grade two. Its song is a repetitive ditty, albeit a spirited one. It is a trill that accelerates to a flourishing finish, always sounding cheerful. The song is so simple that you can hear any variations quite easily – each bird will sing two or three different song-types, and switch between them regularly. This has made the simple chaffinch song a favourite among researchers studying songbirds, and we will come back to what it has revealed later. For now, it throws a rhythmic riff into the atmosphere, adding cheer to the chorus, which is swelling by the minute.

Although it is still dark, another voice arrives late on the scene. As is its wont, however, this voice soon dominates; fancifully I wonder if its loudness is a cover-up for the individual's tardiness. It probably isn't, but any individual starting the dawn chorus late is genuinely remiss and a no-show can have consequences, as we will see. To my ear, the song thrush always seems to proclaim rather than actually sing. It doesn't shout, but its loudness, clarity and studied, measured delivery ensure that it is instantly picked out. Uniquely among the dawn chorus singers, it sings a short phrase that is then repeated two, three or four times, much as a human language teacher might repeat a new word, slowly and almost slurred, to a less than attentive student. When introducing the song thrush, it is almost impossible not to quote the famous words of Browning in his *Home Thoughts from Abroad*. The erudition and rhythm of the poem suit the bird perfectly:

> *That's the wise thrush; he sings each song twice over,*
> *Lest you should think he never could recapture*
> *The first fine careless rapture!*

By now the fine folk of Cerne Abbas were tuning in with a degree of satisfaction. Their ears would not be gorged with the searing beauty of birdsong on such a morning, but at least the sequence so far was clear enough. Blackbird – robin – carrion crow – wren – chaffinch – song thrush. So, indeed, was perhaps the loveliest aspect of any dawn chorus, the swelling of the sound. A dawn chorus starts, as we have heard, with the odd quiet whisper, breaking the dominant silence. A few voices seem to be speaking out of turn, but not yet over each other. But the turning point is when two voices sing at the same time, and from then on there are no gaps. Voices arise until there is a quiet babble, then, almost imperceptibly, a noisy babble. Gradually there is a cacophony, even on a cold, windy morning. The rising volume of the dawn chorus is always slow, never sudden, and the same applies to its end. Twenty minutes or so from the start, the first singers have already finished. At the same moment you realise the dawn chorus has passed its zenith, and that, unexpectedly, it has become noticeably lighter around you. The birdsong doesn't stop – at the height of the breeding season it can go on all day – but its wave has crashed and died down. The growing light dampens the song.

The growing light, though, is also the cue for the later risers. A great tit makes a few desultory songs, but now, in May, its song output is already declining. This individual could well be an unpaired bird, singing a few hopeful songs while there is a little first-light hope. It utters a few 'tee-cher, tee-cher' double-notes; these sound a little like a foot pump and indeed, I once coaxed one into song simply by pumping up my tyres in front of the house. A blue tit, too, is stirred into action; much of the time you cannot tell a song from a scold with a blue tit, but this one is tuneful enough.

By the time the chiffchaff starts, the early risers have sung their bit and have fallen silent. The dawn chorus isn't tired, but to me, it sounds less rich and a little higher-pitched and tinny by this stage. But the chiffchaff injects some extra energy. This sprite, which is blue tit-sized but quite dull olive-green all over, is named after its song, a sequence of 'chiff' and 'chaff' notes repeated in long sequences in metronomic rhythm. The great tit's song is similar, but it usually repeats its cheerful 'tee-cher' notes just three to six times in each sequence, while the chiffchaff can 'chiff' and 'chaff' twenty or more times, and each time the singing bird wags its tail and moves its body from side to side. Its dogged persistence means that the chiffchaff can create its own atmosphere, and to most birdwatchers it is everything about newness, freshness and indomitable enthusiasm. The song is very simple, more department-store kitsch than Albert Hall, but since the chiffchaff is the earliest common migrant to arrive in Britain from the south, in March, its song for me evokes optimism and joy. Even if I play the song on the computer, I am transported to a scene of hazel and willow blossom on a sunny day in March. By now, in the noisy month of May, the powerful message of the coming spring is old news.

As an aside, the chiffchaff's scientific name is a subtle but delightful play on its song. Its genus name *Phylloscopus*, shared with many similar small green warblers, means 'examiner of leaves', referring to these birds' arboreal habitat, less confined to the low vegetation than other kinds of warblers. However, the word *collybita* is a corruption of the Greek word *kollubistes* meaning 'money changer', and your first thought when you hear this is bemusement. However, Greek money changers used piles of coins, each pile with a different denomination and therefore size and shape. Putting a coin on one pile could sound like 'chiff' and a more expensive coin like 'chaff' – you get the idea. It is an example of the imagination and knowledge hidden in many a name in italics that we avoid.

I was hoping for other birds to add their voices to the chorus, but the dawn chorus is usually no more than a twenty-minute burst and it was now all but spent. The light had beaten the cold and wind and we could now see properly around us. The sluggish river glinted as the water flowed down through the village in stately fashion. The fleshy green leaves of the lords-and-ladies began to shine at our feet, the ivy-draped trunks of the path-side trees began to show the details of their shape and the brilliance of the oak-leaf canopy crept out of the shadow. The wind even began to drop a little, enough to hear the river eddies giggle. The birds were still singing, but the once crowded trading floor of the dawn chorus had been replaced by a few loud voices, like a street market winding down.

The Cerne Abbas dawn chorus was by no means a classic, but at least it happened. The cold always reduces the amount of singing, and the wind affected our appreciation of it. The villagers, however, turned out to be enraptured – albeit particularly during the post-chorus full English breakfast. I was amazed at how excited the people were, and the fact that they invariably felt that the experience was worth it. Some of that might have been down to personal satisfaction, a similar sort that you might experience after a run or some other physical exertion. But honestly, if a few bird songs heard in the freezing cold darkness can thrill people, one can only conclude that the chorus holds a special place in people's view of nature.

Yet the dawn chorus itself is also a mysterious phenomenon – the true dawn chorus, that is, not the post-chorus early morning birdsong that continues, like coffee-shop conversation, all day long. If you ask the dawn chorus question baldly, you might appreciate the puzzle: why do so many birds sing at the same time in the dark?

One quick answer goes in here: it isn't to cheer themselves up.

The sentiment of birds being joyful at the approach of the new day has percolated into our culture so that few people doubt its wisdom or truth. I have asked many people over the years why they think birds sing at dawn, and they almost always use phrases such as 'welcoming the day' and 'expressing joy'. If you ask the general question as to why birds sing, you almost invariably hear something like 'because they are happy'.

Poets and writers down the ages reflect this universal, positive view of the pulse of song at dawn. Here's an example you might not have read before, written in the 1940s by Gertrude Tooley Buckingham, an American (so the robin she describes is actually a blackbird-sized bird called the American robin):

I heard the sweet voice of a robin,
High up in the maple tree,
Joyously singing his happy song,
To his feathered mate, in glee!...

Eloquent though these words are, and universal though the sentiment is among people at large, it is fundamentally wrong to think that birds sing to express happiness, or any emotion like it. Despite a few dissenting voices, who maintain that birds get pleasure from singing (the jazz musician and author David Rothenberg wrote a whole book trying to persuade people of this and failed abjectly), the overwhelming majority of researchers and scientists are convinced otherwise. And at this dawn in Cerne Abbas, especially, there would seem to be little pleasure among the birds. Their singing would seem to be more of a compulsion than anything else.

You could turn the question around and ask: what makes us suppose that birds welcome the day and express joy at the increase in light intensity? Is it simply because that is what we humans would do? We are accustomed to singing for joy and have done it throughout history. So it is easy to understand why humanity would draw this conclusion about the dawn chorus.

Our ancestors, those who lived outdoors, would have welcomed daybreak with intense pleasure. Dawn, for them, would represent relief at the end of the night. Even here in Britain, most animals of real or perceived danger to humans would have been nocturnal. Bears, wolves and, going back still further, terrifying beasts such as cave lions and sabre-toothed cats would have struck mostly by night and, later on, so would human enemies. Then, darkness would bring menace and we are still habitually afraid of it. No wonder we greet the slightest trace of light in the sky with delight and joy. The similarly noisy response from waking birds, apparently ecstatic at the same time, would have convinced our distant ancestors that birds were feeling the same emotions.

Another potentially powerful unifying emotion would be delight at the onset of spring. As we all know, the dawn chorus is at its height in spring, that same savoured time when all people could begin to look forward to a more generous environment, when the earth warmed up and both vegetation and animals would be more prolific.

However deep the connection, though, it seems to be entirely false. Birds don't sing because they are happy and they don't sing to welcome anything. Here we have, once again, a deep misconception about the birds with which we share our gardens, woods and airwaves. The function and practice of singing is quite different from what we might think.

It only takes a moment to puncture our assumptions with some basic observations. After all, do birds not quieten down after the spring? Beyond June the woodlands and hedgerows largely fall silent, in July and August there is nothing. In September and October a few voices start up again, notably robins, and it isn't until February and March that the dawn chorus, and daytime singing, begins again.

Are birds only joyful in the spring, then? Are they unable to welcome the day at the heart of winter? If song is an expression of happiness, or even of feeling buoyant, why aren't the wild airwaves busy all year round? Is there a moratorium on joy outside of the spring and early summer?

Here's another observation that anyone can make that casts doubt on whether singing is about joy. Look carefully at exactly which birds sing in your garden, or elsewhere. When it's possible to distinguish the sexes, which one do you see singing? I am writing this in the spring and several blackbirds adorn the roofs of our street, singing gorgeously almost all day. They are all black with yellow bills, adult males. Similarly, I have watched the chaffinches singing in our nearest wood and they are all colourful, with bright pink breasts and with heads a neat shade of blue-grey. I have never once seen a brown female blackbird singing, nor have I noticed the plain mouse-brown female chaffinch singing. All the singing yellowhammers I have seen have been brilliant buttery yellow, particularly around the head, and all my singing blackcaps have had black caps, not brown. My observations are only a snapshot of what goes on, but never have I seen any females of these species sing.

If birds sang for joy and with delight, it should be for both sexes to proclaim their feelings. Yet in all the bird species I have mentioned, it is the males that sing and worldwide there is a male majority devoted to the task, if not entirely a monopoly. Ask the scientists and you find out that this is true for most species, but not all, throughout the world. If this is the case, then surely song has a specific biological function that is carried out by males?

At this point, I should make clear that not all the sounds made by birds are classed as songs. This is important to recognise to understand what follows. Your average songbird makes a variety of sounds, many of which are just of one or two syllables, not more than snatches of sounds, analogous to single words. These sounds may be uttered as contact calls to keep a flock together ('hello'), or they might be an alarm reaction ('look out!'). They are used in a variety of contexts and, importantly, many are uttered all year round. Ornithologists describe a broad range of these utterances as 'calls'.

Songs are something different, and the main distinction is that songs are more elaborate. The duration of a call will tend to be less than a second, whereas a song lasts longer, sometimes much longer. To use the analogy of human speech: if a call is a word or a syllable, a song is a phrase, a sentence, or even a poem. As it happens, these distinctions work for most of the time, but not all of it: for example, 'cuck-oo' is best regarded as a song, while the angry trill of a blue tit, while it might last several seconds, is functionally a call. Scientists will tell you that there is no clear dividing line between a call and a song, which is true, but it is important that we treat a bird's song as something entirely distinct from its calls, in a category of its own.

The true difference between songs and calls can be found in their respective functions. Calls may be uttered in many different circumstances, but they tend to be reactive to a stimulus such as danger, being lost or even keeping contact with a mate or parent. Songs are different. Although they can be uttered in retaliation to a song from another bird, they are not usually a response to something, but an attempt to assert something.

That cause is territory, and it is the primary function of birdsong. Song has other functions, including mate attraction, as we will see, but when birds begin to sing early in the season, often from January onwards, it is all about territory. Many species of British songbirds, including blue and great tits, finches and buntings, live in flocks in the non-breeding season and are silent other than making alarm calls and contact calls. As a prelude to breeding, however, they must acquire a patch of ground to call their own, a territory from which others of the same species (but not usually different species) are evicted. This is necessary for spacing pairs out, ensuring that neighbours don't unnecessarily interfere with the breeding attempt. A territory might contain exclusive feeding resources for a pair and family at first, but normally, in the exhausting phase of feeding the young, barriers break down and everybody trespasses. However, most birds will aim to forage within or close to their own patch of ground.

It would be difficult to over-emphasise the importance of a territory to a male bird. If he cannot secure one, his breeding season is doomed to failure. If he cannot secure a good one, his breeding season is in the balance. His territory is an extension of his fitness as an individual because the best quality birds secure the best available territories. The quality of his territory lays a male bird bare. So, if anybody needed any reason to cast aside the notion that bird songs were signals of happiness, contentment or joy, then this cast-iron relationship with territory would surely convince.

If you have watched birds for any length of time and listened carefully to them singing, you have probably witnessed some impressive vocal confrontations. These are the times when birds don't seem to be singing, but shouting at each other. This happens during the dawn chorus when birds seem to be answering each other, but it is more obvious at other times of day when a vocal spat can dominate the soundscape. A bird sings, then a rival of the same species answers; the answer is a little quicker than might be polite, and louder than necessary. The first singer might then intensify its song, and sometimes the phrases take on a curious, strained quality. The phrases go back and forth, with each bird beginning its response before the other has finished singing, stepping on each other's toes vocally. Such song duels, or 'counter-singing' matches, often end in a physical chase and abrupt silence. Everything about them shouts of rivalry, of confrontation, of 'song-rage'. Here there is no peace or joyfulness; more than a hint of bullishness, perhaps, but an inflammatory bullishness.

I have witnessed these song duels many times myself, and they are always compelling. By far my favourite was between two nightingales that lived, as it happened, in southern Turkey, where I had gone on holiday with a sideline in recording the bird sounds. These hot-blooded rivals were neighbours, occupying bushes on either side of a dusty road. Even in mid-morning, when I arrived, their songs were loud and piercing. Every few moments they would counter-sing and become, apparently, more flustered. But of course, being nightingales, their vocal skirmish was expressed in the most glorious exchanges of song: loud, varied, inventive, egging each other on towards more wondrous creativity. It was like listening to an argument between poetry fans, each yelling ever more complex and noisy extracts from Shakespeare. Eventually one individual clearly became bored with the war of words and delivered the *coup de grâce*. It rattled off a single trill that alone lasted for thirteen full seconds. To me, listening on the headphones, it sounded like a stuck record, and I nearly ruined the recording by laughing out loud. It was assassination by trill. The rival fell silent, humbled.

Let's return, though, to the dawn chorus itself and ask a few questions. The first one is: are all these voices really males? The answer is yes. That fabulous, musical, sweet sounding stream of loveliness is not a choir of angelic voices, but instead the sound of testosterone-fuelled rival males all yelling at the same time. The voices are far from soothing in nature, but instead full of assertiveness mixed with a sprinkling of menace. How we've got the dawn chorus wrong!

Once we accept that the dawn chorus is sung for territorial proclamation and defence – and incidentally, each male is only singing to his own species – we are only part way towards understanding why the phenomenon happens. This only tells us what song is for, not why it happens at dawn. A bird could launch a breach of a rival's territory at any time of day, not just at dawn. And besides, birds don't confine their song output to dawn, they just give a brief burst before the sun rises, in the dark.

Why, then, do they sing so frenziedly before dawn? You might be surprised to hear that, while hypotheses abound, there is no definitive answer and no single explanation. Most of the suggestions make perfect sense. For example, it has long been known that the hours at the end of the night are characterised by favourable atmospheric conditions. There is a perfect storm of temperature, humidity and reduced wind turbulence that helps to propagate acoustic signals. In woodland, a ceiling of trapped air apparently creates a sort of tunnel through which sound travels well. And, of course, before dawn is the quietest time of day as far as competing ambient sounds are concerned, those from other animals and those from the human world. Scientists have concluded that a sound emitted at a typical bird's pitch will be transmitted twenty times more efficiently in typical dawn conditions than it would be at midday. Thus speaks the acoustic transmission hypothesis.

It makes sense to sing when you can be most clearly heard, but this hypothesis doesn't feel like a complete answer. After all, in the early breeding season birds do sing all day, including at midday when they aren't heard to advantage but must still defend territories. This idea also doesn't explain the very intriguing sequence in which birds start to sing, which tends to be very consistent from place to place – robin first, blackbird second, and so on.

The only hypothesis that does partly explain the order is the splendidly named 'inefficient foraging hypothesis'. This maintains that birds sing at the breaking of dawn because, while ambient conditions are ideal for singing, it is still too dark to make foraging worthwhile. (A useful corollary is that, when it is dark, the birds can sing loudly and lustily while their main predators, day-hunting birds of prey, are not yet active – although of course the birds won't be safe from owls, which are at the end of their hunting shifts.) This may at least begin to suggest why there is a sequence of species to the dawn chorus that is stable in a single place, wherever you are in the world. In Britain, the large-eyed robin and blackbird provide the first voices you hear, and they have usually tailed off by the time birds such as great tits or chiffchaffs begin. Their large eyes help robins and blackbirds begin to sing earlier and then forage earlier. Many British dawn-chorus singers feed on insects, which tend to be well hidden in vegetation, and are therefore difficult to find in the semi-darkness.

In tropical forests, the few studies there suggest that, among true songbirds, the dawn chorus begins earlier for those species that feed higher up in the canopy. This would make sense because it gets lighter in the canopy much earlier than in the middle strata and at ground level, where it can be gloomy all day. This is what you would predict using the inefficient foraging hypothesis.

This hypothesis tells you as much about why birds stop singing just after dawn as it does about why they sing in the first place. Yet another explanation of the dawn chorus, and why birds sing at all at such an hour is that it is effectively a roll call of who is present and correct in any given place. The birds sing in order to re-establish the order of things and to confirm who is present in any given territory. The chorus enables the bird community to establish if any of their species has succumbed overnight, through starvation or illness, and therefore whether there are any new vacancies available. The dawn chorus would be a convenient time for an interloper to make a move. Any bird listening in, either male or female, would have a dawn update of the comings and goings of the neighbours.

There is some evidence, indeed, that females play their part at the time of the dawn chorus. They don't sing, but studies of great tits show that the female may interact with her mate by calling from inside the nest-hole. Just before and during egg-laying, the female utters a very soft, female-specific 'quiet call' which is only ever heard in this context.

In recent years, the potential reaction of females to the dawn chorus has produced a quite different additional hypothesis as to why birds sing just before daylight. The idea was first put forward when scientists measured the peak output of dawn-chorus song in great tits. They found that the males sang most vehemently during the period that the females were producing eggs (the fertile period) and during the egg-laying period itself – the time, indeed, when the females are quietly calling to them. Since great tits take up territories and pair up long before this happens, the inference is that the peak in dawn-chorus singing is related to the state of the female, not the male. During this time of year, as soon as the female gets up in the morning and leaves the nest-hole, the male great tit immediately copulates with her, and then his song output reduces sharply. At this moment female great tits are at their most fertile, and it suggests that a paired-up great tit would be well advised to sing with as much gusto as he can when the mate is broody. If not, the female might seek to mate with a different male worth his salt.

Another study found that higher-quality male blue tits, as measured by age and experience, began to sing earlier at dawn than less experienced birds. This would be a gift to an unpaired female, or one that was seeking to mate outside the pair bond (more of this in chapter 3). By listening for early birds among her own species, she would quickly be guided to a high-quality male.

All the hypotheses suggested for the dawn chorus are a far cry from the fanciful suggestion that birds sing to welcome the dawn, or that they are expressing joy and happiness. If we are being honest, and look into the real meaning of birdsong, you might even say that, far from expressing joy, the dawn chorus is no more than competitive shouting. The birds are fundamentally singing against each other and, while their emotional state could be described as bullish, it isn't joyful.

In the end, to describe the dawn chorus as a bunch of happy creatures worshipping the arrival of the sun is to demean a far more intriguing phenomenon, and to bring something amazing down to the low common denominator of human feelings. Of course, it is wonderful for us to listen to the dawn chorus and let the joy and spring thrill percolate through our emotions. There are few better experiences of the natural world. But if we truly understand what is going on, in all its intrigue, we can actually find the discovery still more rewarding.

CHAPTER TWO

Take Your Partners

In the suburb of south-west London where I grew up, we used to have a good population of song thrushes. They would carry the soundtrack of the breeding season, starting up even as early as a quiet November evening, and continuing through the coldest weather, vocally insisting that spring was inevitable. Frosty mornings in January and February were made bearable by hearing their loud, ringing chorus, the wild calling before the school day. In spring there were other voices, but still, the thrushes dominated, by virtue of their clear, slow, repetitive song elements and strength of sound.

I can remember one year being dominated by a particular individual. The memory lingered partly because he held territory in our garden, but also partly because, after everything else had quietened down, even during the heat of a summer day, this bird would carry on, relentlessly singing.

But the main reason I so clearly remember this individual is that he was a simply dreadful singer. Even I could tell this, so what the female thrushes made of him, one can only guess – females can be cruel when they perceive a male's inadequacies. I called him Bob, after Bob Dylan, whom I considered to be the worst and most overrated singer of all time. Actually, this is harsh on Dylan because, despite his awful voice, he is at least inventive. Bob the song thrush was sorely lacking in everything – delivery, skill, repertoire.

The song of the song thrush is well studied, and the individuals analysed have been shown to sing many different 'phrases' or units: the fewest recorded for one individual is 109 songs and the most is 219. The trouble with Bob, though, was that he only ever seemed to sing two or three types of very simple phrase, and these he used to repeat endlessly. It became a joke. The peace of the back garden was constantly broken by Bob's 'p-peep...p-peep...p-peep' phrase, to which he returned after a few dalliances with other notes. He began in January and was still going deep into July, 'p-peeping Bob'. To misquote Browning:

That's the dire thrush, he sings one song over and over.

It was pretty obvious that singing badly was never going to win Bob any favours, and so it proved. The very fact that he sang endlessly in the summer is proof, if it were needed, that he did not attract a mate. As soon as song thrushes pair up and begin to breed, the male stops singing during the day, confining his efforts to the dawn chorus (and perhaps the evening chorus). He will sing again after bringing up the first brood, to re-establish his territory in June for a second breeding attempt, but is silent for much of the spring (this, incidentally, is why people may think song thrushes are rarer than they are). Unless, of course, like Bob, he will sing forever and a day, in hope.

Sad though Bob's story is – and his misfortune continued the next season, when he was no better – it was this individual that demonstrated to me for the first time that some birds are better singers than others. As I learned more bird songs in my youth and began to listen more and more carefully, I soon discovered that many of the individuals that I was hearing came up with unusual phrases that seemed unique to them. There was a chaffinch that ended with two final flourishes instead of the usual one, there was a great tit that seemed to be drunk and slur its phrases. There was a blackbird in our garden that imitated the sound of a Trimphone – remember those? Everywhere, if you listened carefully enough, you could pick up individual variations. It was clear that every singing male had a repertoire.

When nowadays I tell people that individual birds have repertoires, they usually sigh deeply and decide that they will never crack the code of birdsong: if each individual blackbird sounds different, how can you ever learn to recognise a blackbird? Of course, the truth is that variations in a repertoire must be made in the context of the species itself, so they must be subtle and fit in with the 'rules' of the song. If a blackbird was so inventive that its song no longer sounded like a blackbird, then an enhanced repertoire would be self-defeating – the bird wouldn't be understood. So birds can only vary the template that their species has set. The reality is that you can only recognise individual birds in the field by listening to them very carefully indeed, and often it is impossible.

What advantage does a bird gain from having a wide repertoire? Quite a lot, it turns out. A wide repertoire is a good opening gambit to impress a listening female or intimidate a rival. Birdsong is nothing if not self-advertisement. A good song makes an impression on the entire listening community, and there are only so many ways you can stand out from the crowd. You can sing more often than your rivals, you can sing more loudly and aggressively, or you can demonstrate your repertoire.

Definitive proof that size matters in song repertoire has come from a number of studies, including one on sedge warblers. These are zestful sprites that are summer visitors to Britain, arriving in April and settling into the margins of marshes, usually where thorny scrub grows. Their song is energetic, vigorous and continuous, the phrases lasting longer than those of most birds with the result that the song sounds like a fast, swearing monologue. As a sedge warbler sings, it often begins well down and concealed in a bush, climbing up little by little until the bird can be seen just below the top, having crept up without breaking its stream of babbling 'conversation'. It is quite a performance, but these showy singers have another trick up their sleeve. Every so often they launch into the air as if fired from a cannon, then beat their wings slowly as they float back down to a concealed perch, all the while continuing to sing.

To study the link between repertoire and breeding success, Clive Catchpole, a researcher at Royal Holloway College, University of London, recorded the songs of three sedge warblers that were colour-marked and therefore individually recognisable. He and his co-workers analysed the songs in the laboratory by using a sound spectrograph, which converts sound into a graph of the pitch against time. The pattern produced is distinctive, making it easy to divide a sedge warbler's stream of sound into its component parts and, therefore, measure its repertoire (the same apparatus could analyse a human being's vocabulary). In the wild state, Catchpole was able to prove that the male sedge warbler with the richest repertoire paired up more quickly than his colleagues and the sedge warbler with the simplest song was the last to obtain a mate. Catchpole then transferred the experiment away from the wild state, where a male's territory quality and other factors could have played a part, and caught a number of female sedge warblers as soon as they arrived from their migration. Here he subjected them to recordings of the three individuals mentioned above, and controlled all the variables such as song output and rhythm, leaving only repertoire size as a variable. This time he measured the females' responses in terms of their soliciting displays in their cage to the broadcast songs. Once again, the richest and most varied repertoire elicited the strongest response.

Since then, repertoire size has been proven to be a constant factor in female choice across many species. Song has proven to be a factor in sexual selection. Not only does birdsong allow a male to claim a territory, at the same time it can also act as a measure of male quality, which enables females potentially to choose the best partners simply by their voices.

It turns out that a song really is an advertisement; it is a vocal profile on the dating airwaves. A male can attract a female before even meeting her, and a female can listen discreetly to a male without necessarily making any visual assessment of his qualities.

What qualities, though, might be present in a male with a wider repertoire, attributes that make it a good idea for the females to assess male song? In recent years, scientists have found an aspect of breeding in which sedge warblers with more varied songs are superior. Males with better songs show a generally higher provisioning rate for their chicks, so their nestlings and fledglings grow more quickly. How being a diligent father is linked to having a better song repertoire is not known.

Although repertoire is important, especially among species with very complex songs, the sheer rate of song output is also a good measure for the discerning female bird. Many of us have witnessed how some birds just seem to sing more energetically than others, and it seems that the very simple act of being in the audience and monitoring how much a bird sings can help a female make a choice, at least about whom to visit with a view to pairing up. A case study of willow warblers in Sweden showed that males that sang more often per unit of time acquired their mates more quickly than singers with lower output.

But why do females prefer males with a higher output? It might be simply that such males sing more because they are physically healthier, and thus better specimens that will contribute good genes. Or maybe something else is at play here?

The researchers suspected that, in the case of willow warblers, territory quality might be related to song output. So they measured the number of songs a male delivered per unit time and found that it varied consistently between individually marked males. They then removed the territory holders and allowed in other males to replace them and what did they find? The new males sang at a similar rate to the previous incumbent, and this varied across the board. The male holding territory sang at a rate apparently determined by the territory.

This made sense. Willow warblers are frisky singers. Their lovely, very gentle cadence, a descending scale of sweet nothings that seems to blow away on the breeze, comes and goes quickly. A bird such as a yellowhammer will plonk itself on a high perch, all yellow bling and confidence, and sing there until kingdom come – such that you sometimes wish it would just shut up for a moment. The willow warbler rarely performs like this; it will sing a few phrases, but these will melt away into a quiet patch, before you pick up the odd phrase again. The willow warbler doesn't have a complex song, and it doesn't seem to work particularly hard at the song it does have.

So if you watch willow warblers in the low scrub (often birch, as well as willow) where they breed, they sometimes give the impression of singing on the hop. Indeed, the researchers suggested that it was the juxtaposition of song and feeding on a willow warbler's timetable that made a difference. A male willow warbler singing from a rich territory providing ample food should be able to spend less time foraging, and more time singing than a male with a less bounteous territory. It stands to reason that, if a male is compelled to spend more time searching for food to sustain himself, he will be able to devote himself less to singing.

In other words, willow warbler song contains a coded message about territory quality that females should listen to keenly. A territory with better food supplies close at hand will make it easier to feed the resulting chicks, and enhance a female's own chances of a successful breeding attempt.

This is not the only instance when an aspect of a male's song may send an unintended code. A delightfully amusing instance of giving away information concerns our old friend the sedge warbler. As I mentioned earlier, the sedge warbler adds an intermittent visual feast to its song by lifting into the air on a brief, not very spectacular songflight. A recent study has shown that the frequency of songflights is negatively correlated to the male's state of health, as expressed by the amount of parasite infestation on the plumage. Birds routinely have to cope with feather lice and other parasites, and when they suffer a heavy infestation it invariably affects their overall perkiness. Since resistance to parasites has been shown to be partly inherited, it would be very useful for a putative partner to tell easily how a particular male was faring. A high songflight frequency and large song repertoire both correlate with low parasite infestation.

Before leaving the subject of song and mate choice, it is well worth asking whether there is any evidence that the female birds actually use the information with which they are provided. Do they go around the area checking out the songs of different males? It is difficult to monitor this, but all the evidence suggests that they do exactly that, at least they do to a limited extent. Female pied flycatchers, for example, visit four males' territories on average, and some other birds select six. However, too much checking wastes valuable time that could be spent getting ahead in the breeding cycle, especially if other females are on the prowl simultaneously. So it seems that reality takes over, and most females choose quickly rather than hedging their bets.

So we have established that song is a vital tool in the process of acquiring a territory and attracting a mate. However, song is alarmingly a tool of disclosure. Open your mouth to sing and subliminal information pours out, on your overall quality, your state of health, your ability to be a good father. In many cases, the song gives your age away, especially in your first year when you are trying it out for the first time and it is rough around the edges. If birds could be embarrassed, singing might give them red faces.

And speaking of red faces, what about a bird's appearance apart from its song? Might that give any clues to the choosing sex, the female, as to whether a partner is suitable? In recent years, mate choice driven by sexual selection has been a hot topic among biologists, and there is now a cornucopia of research on the subject. We now know the intimate decision-making processes of birds such as house sparrows, blue and great tits, barn owls and a host of species many of us have never heard of. We'll probably work it out in retrospect for dodos one day. The irony in all this is that we still barely understand what closes the deal for humankind, especially since this varies across the world. I personally understand the wooing characteristics of our garden blue tits better than the mysterious underlying chemistry that brought me and my wife Carolyn together. I have a good idea of the first but all I can do is revel in the joy of the second.

Red faces are, by the way, sometimes a cue, and this is the case, for example, in a much-loved common bird, the yellowhammer. This was once an everyday farmland staple in our islands, so much so that just about everybody knew its dry rattling phrase. In my youth, I used to have discussions with all kinds of people about the most appropriate mnemonic for the song – 'A little bit of bread and no cheese' was the best-known, but unwieldy one. But nowadays most people have never seen a yellowhammer, let alone heard one; its population declined by 55 per cent between 1970 and 2010 and it disappeared from more than a fifth of its former range. It is one of our few real yellow birds, adorning the tops of hedges 'with yellow breast and head of solid gold' as the poet John Clare put it.

If you take a close look at a male yellowhammer in spring, this 'solid gold' head is exactly what you notice – you cannot miss it. It is also obvious that different males have different degrees of yellow colouration and some have streakier head patterns than others and some are simply plainer. Yellow is an awkward pigment for birds to synthesise because the chemicals that produce it, carotenoids, are not standard by-products of a bird's internal chemistry – as are melanins, for example. Instead, carotenoids must be obtained from the environment, which takes effort. Carotenoids are found in leaves and, for most birds, the best way to obtain them is indirectly by eating leaf-consuming caterpillars and other insects. Experiments with captive birds have shown that the more carotenoids a bird eats, the brighter its plumage becomes.

In the case of the yellowhammer, it seems that the females take the yellow colour for granted. It is any redness in the plumage, not just on the head but on the body too, which makes their juices flow. Red also comes from carotenoids and is apparently harder to synthesise than yellow. The brighter a male's colour is, the better impression this makes on females.

The male yellowhammer is perhaps typical of birds in that he betrays a battery of variable features in the breeding season, each of which speaks honestly about his quality, whether he might benefit from showing off these features or not. The song repertoire of this bird has been studied too and, as in the case of sedge warblers, a more varied song is also a factor in sexual selection. So is the poor bird's parasite load. In the breeding season, individual birds cannot escape their failings; they are there for all to see and hear. In this world there are many failures, and it seems that failed birds know all too well their failings and can barely mitigate against them. In our human lives, we have softened the sharp edges of mate choice; immediate failure, at least, doesn't need to be so devastating.

Back to yellow birds, though, and we find, once again, that what we thought might be simple is very far from it. It turns out that the carotenoids send a much deeper message than a stark choice of high hormone quality. They offer a glimpse into a bird's environment, its past, and its potential future.

An elegant study on great tits in Norway showed, quite amazingly, that there is a relationship between the yellow colour on the breast and the habitat in which the bird was born. Chicks reared in deciduous woodland turn out to have yellower breasts than birds born in coniferous woodland. Conifer forest always has a lower population of great tits than equivalent broad-leaved trees, and it is clear that the birds prefer the latter because it provides more abundant food. The inference is that chicks reared in broad-leaved woodland would have been fed either on more caterpillars than their conifer equivalents (which is what the scientist concluded) or on caterpillars with proportionally more carotenoids in their body tissues; either way, the abundance of food filters down into the carotenoid intake and into the intensity of the yellow plumage. When the researchers switched eggs from deciduous nest-sites to coniferous nest-sites and vice versa, to mitigate against genetic and other differences, the chicks in the deciduous sites still turned out brighter.

Is, though, a difference in breast-colour intensity of use to another bird except to say 'oh hello, I'm from a conifer forest as well'? The answer to this question could well be yes. If a bird from a conifer forest is duller, that could betray the fact that it is from marginal habitat stock – from the rough side of town, if you like. A bird selecting a mate might be put off.

More importantly, though, the yellowness of its plumage might also indicate something about the bird itself, not just its origin. After all, once a great tit has fledged it will move away from its parents' territory and fend for itself. It is independent before it moults in the late summer, so much of what it synthesises in pigment then will be down to it as an individual. If it eats lots of caterpillars in the late summer, it should acquire vivid yellow. Its colour will be an indirect measure of how good it is at finding caterpillars.

And here's an intriguing thought – when do tits really need to be good at finding caterpillars? It is, of course, when they are feeding their young in the late spring. At such times they need to collect 400–500 caterpillars during the daylight hours for two weeks in succession. Now put two and two together and what measure could you use to monitor a potential partner's ability to look after their youngsters well? It's the brightness of their yellow plumage, of course. There could be a link between a bird's colour and its ability to find food for the young. Humans would kill for genuine information on a potential partner's skills like that!

In fact, several studies have unveiled links between a male attribute and his eventual quality as a father, expressed as being good at bringing food. This is very important, because when females are occupied with brooding chicks that have only just hatched, they need to be able to rely on a male's food provision. Several studies have found links of this kind. Conny Bartsch and her team from the Free University of Berlin found that male nightingales with better, more ordered songs are advertising their qualities as fathers, and they back this up with more frequent feeding visits to the chicks.

It is never as simple as this, however. In a remarkable study, blackcaps were shown to advertise the quality of their territory by means of their singing rate – if they sang 160–180 times an hour, a female would have dense vegetation in which to place the nest, offering enhanced protection from predators. On the other hand, males singing in the range of 80–100 songs an hour could offer less thick vegetation, but compensated for this by offering more help with incubation and chick feeding.

In such cases, the choosing sex, almost always the female, faces a trade-off. Does it simply go for the best genes, and expect to do most of the chick rearing by itself? Or does it lower its standards slightly and pair up with a mate that is likely to be a reliable and devoted parent? It is an intriguing choice.

Before we leave the subject of carotenoid pigmentation, we should note that it isn't confined to the feathers. Quite a number of birds have yellow or red bills, and these too can be badges of excellence. In the case of the blackbird, the female has a somewhat grubby brownish-yellow bill, but the male's bill is bright yellow and often tinged orange. In this case it is also a measure of good health, the inference being that, if a bird can put energy (in the form of converting food into carotenoid pigments) into the colour of its bill, it can't be wearing itself out fighting any kind of infection. Interestingly, the bill colour changes according to how the bird is doing, with a slight lag period, so a potential mate can monitor a blackbird's recent state of health.

The relationship is actually more complicated than that. In a recent study, it was found that it is the relative numbers of certain parasites that are expressed in the bill colour of a blackbird, not whether there are large numbers of them. Feather-louse infestations don't affect the bill colour, but certain blood parasites do.

Carotenoid pigments are not the only expressions of a songbird's quality that can be monitored by the opposite sex. In recent years, literally dozens of different attributes have been found to play a part in sexual selection. It is really a case of different bird, different 'badge'. Some of these badges are well hidden to us, such as subtle differences in colouration, including some in the ultraviolet spectrum.

One of the first to be discovered concerned swallows, and the fact that females are choosy about the length of tails in their males. This, in a way, is hardly surprising, since tail length is a good way to tell the males, which have the longer tail-streamers, apart from the females in the field. And sure enough, the longer the tail, the more attractive the male. However, long length isn't any good if the tail-streamers are not symmetrical, because that represents, perhaps, a poor attempt to make the streamers long without having the internal resources to cope with doing so. Streamers act as a flying impediment, so a male with long streamers is declaring: 'I can fly well even with these ridiculous things'. And the females are duly convinced of the long-tailed males' quality.

The way scientists discovered this is a story in itself. Anders Møller and his team from the University of Aarhus, Denmark, carefully watched various male swallows at a colony so that they could monitor their pairing and productivity. They found that the longest-tailed males took about three days to acquire a partner once they began courting, while males with normal-length tails took eight days and those with the shortest tails took up to twelve days; the females, for their part, clearly took some time over choosing. Meanwhile, Møller also found that the longest-tailed individuals, pairing sooner in the season, were more likely to father two successive broods. As a result, they produced an average of eight chicks, which is twice as many as the shortest-tailed males.

How, though, did they prove that it was tail length that was a key to mate choice? It was simple enough. They captured male swallows during the pairing-up process and literally trimmed the tail feathers of the longer-tailed individuals, cutting them off in their prime from 10cm long to 8cm (from 3.9 to 3.15in). Then they attached the cut-off tail-streamer tips to the swallows whose tails had been shortest at the start of the experiment, making them as long as 12cm (4.74in). For control purposes, they cut the tails of some other individuals in all categories, only to glue them right back on again to leave the length unaffected but the process the same. Sure enough, the mating success of the individuals with trimmed tails plummeted, and that of the controls remained the same. Those males with artificially elongated tails suddenly found themselves, wide-eyed, at the centre of female attention. All their wildest dreams must have been fulfilled.

The females didn't know it, but those artificially enhanced individuals were a scam, and it so happened that their new tails found them out. Møller noted that their flying was impeded, perhaps because they had tails 'literally' glued onto them! making them rubbish at catching insects and quite unable to bring in any decent-sized morsels for their young. It goes to show that the long tail is a genuine mark of a male's quality.

Another well-known 'badge' is the black throat-patch of the house sparrow – a marking only present in males. The size of the black bib varies between individuals and those with bibs with an area of avoid measurement being split over line 4 or 5 cm^2 (0.6 or 0.8in^2) are dominant over their smaller-bibbed colleagues. Bib size is a big deal. Fundamentally it is a signal of male strength and males with larger black bibs have been shown, as with long-tailed swallows, to be more attractive to females. In the complex world of bird behaviour, good badges always signal something of practical use, and in the case of sparrows this would seem to be territory quality. In one study it was found that larger-bibbed males held territories with more potential breeding sites and, not only that, the sites were of better quality – secure crevices rather than branches or eaves. Hole nests protected the young more effectively, so by pairing up with a male with a large bib, the female gave the best chance to her young.

Bib size and what goes along with it is not something you can fake as a sparrow. As mentioned above, it is a measure of male strength. In experiments in which subordinate male sparrows' badges were enlarged by dyes, the birds constantly found themselves challenged by the dominant birds with genuinely large black patches. The dyed birds always lost these challenges and found themselves at the bottom of the hierarchy where they belonged.

There is, in fact, evidence that house sparrows know only too well where they belong in society. Those with large bibs more readily take part in communal chases, for example. Anybody who has sparrows in the garden will have witnessed the bawdy chases that take place in the spring. In some ways, they look like the gang rapes for which mallard drakes are notorious, but the female sparrow holds her own and the noisy spectacle, with much excitement, cheeping and plumage-ruffling, is mainly ceremonial in nature. But it seems that the sparrows with smaller badges stay on the sidelines, like self-conscious teenagers at their first dance.

However, not all is lost for these lower-quality sparrows, as it turns out. This is because the females in the population also have a pretty good idea of where they stand in the sparrow cattle market. Matteo Griggio and his team at the Konrad Lorenz Institute for Ethology in Vienna have discovered that low-quality female house sparrows (as measured by their poor body weight) have a marked preference for equally low-quality males. Rather than going for the unattainable, they spend their precious energy engaging with the sort of male that is in their league. As a bonus, such a male is likely to devote some extra time to looking after the youngsters.

House sparrows have a bewildering array of characteristics to choose from, it seems, when selecting a mate. Not only is the bib size important, so is the colour of the bill and the length of the leg. It is very likely that, just as in humans, any number of features – and the totality of those features – contribute to who pairs up with whom.

Up to now I have mainly talked about the hard-pressed males, who, if they aren't being body-shamed on the avian equivalent of social media, are being song-shamed as if running the gauntlet of judges on a talent show. Increasingly, though, scientists are finding that female traits are just as variable as male ones (the missing tendency to sing excepted) and are equally useful for obtaining pre-breeding information on a potential partner. In recent years, a number of researchers have been studying information on bird plumage that we humans cannot see – namely, ultraviolet reflectance. Both blue tits and great tits have vivid blue plumage on the top of the head (the cap) and on the wings and tail, and while we cannot easily distinguish individual birds by eye, the ultraviolet reflectance varies enormously.

In female blue tits, it seems, the degree of ultraviolet reflectance from the cap is an indication of how good a mother a particular bird is. Remarkably, females with highly reflectant crown feathers are simply less stressed during the critical part of the breeding season when they are feeding chicks. So, while they don't lay any more eggs than less brightly capped females, they do on average rear more chicks.

For once, therefore, a male blue tit has a pretty good idea of what he can expect from his chosen mate. In the words of researcher Dr Kathryn Arnold, from the University of York: 'a male choosing a brightly coloured female will gain a good mother for his chicks and a less stressed partner.'

It seems that there is nothing that birds of both sexes will not use in their assessment of a potential mate. And nothing seems to be hidden either, from a bird's parasite load, to its hunting ability or its ability to cope with stress. What can we make of bird society? One thing we can say is that it is so overwhelmingly public. In the social life of birds, everybody you meet seems to be able to assess your strengths and weaknesses, and use them in some way to their advantage. In the avian cattle market, the losers are obvious and are sometimes just written off.

As with so much of a bird's life, choosing a partner is competitive and vicious.

CHAPTER THREE

The Breeding Circle

I had been a birdwatcher for more than thirty years before I learnt about the secret. Countless other enthusiasts before me had spent their lives watching their avian neighbours without ever twigging. Even now, with a large body of research for confirmation, few birders have a clue. The general public is completely in the dark. All in all, we have little concept of the weird and wonderful world of bird relationships.

In my own early days as a birdwatcher, my overwhelming experience was of faithful, co-operative songbirds. I would spend hours watching the blue tits in the hole in our apple tree coming back and forth with caterpillars in their bills. The one with matted feathers in its crown would come in, closely followed by an individual with a long broken black line down its breast. After a pause raggedy-crown would return, deliver a caterpillar and slope off, while after a few moments broken-stripe would steal in without breaking its flying stride, but could be identified as it peered out briefly before its departure. Sometimes the same adult would make two consecutive visits, but nothing broke the labour of the two. Over the fortnight or so of chick-rearing, they looked increasingly ragged, but that just confirmed their teamwork was paying off.

It was the same with our blackbirds. There was a coal-black male with his bright yellow bill, and a brown, streakier female with her unclean yellowish bill, and as far as I could tell nothing obvious ever severed the duopoly. True, the male's bill did begin to look less of a clean buttercup-yellow as the season went on, possibly because he was forever holding earthworms between his mandibles. The female was frequently very secretive, but once again, the industry of the birds seemed noble, and their togetherness was not in doubt.

Nobody's togetherness was in doubt, not that of the local jackdaws, the dunnocks and our pair of robins. In the breeding season, I always saw them in twos, in what looked like pairs. Occasionally I would see males fighting, but even that seemed noble – the male protecting his family and territory. When birds bring up their families, against many odds, it is something wholesome to behold. Reputations become spotless.

In the evenings I would scour books and learn about birds. In common with many a child of an introverted nature, I never felt safer or more comfortable than when surrounded by books. They were the conduit to the wider world and a safety net from its reality. These books never wavered from the principle that birds are essentially monogamous. Swans were famously so: they 'mated for life', their morals as spotless as their white plumage. This was probably only quoted because swans lived longer than most other birds. The assumption was that almost any species, given the chance, would be inclined to do the same.

Occasionally I would see individuals behaving outside the norm, or at least outside its respectability. I was an avid visitor to Pen Ponds, two lakes in the centre of Richmond Park in west London, and every so often the mallards there would behave with a lack of decorum. Three or four males would be seen to lose their cool near a female and attempt a group rape; my books told me that sometimes this would lead to the death of the female at the hands of her overenthusiastic suitors.

Songbirds, by contrast, did not seem to partake in anything like this. They gave no real clues. If you watched almost any of them, from a crow to a robin, the assumption was that, once a male had successfully wooed a female with songs and displays, that same, closed unit would go on to reproduce as best it could.

It turns out that this premise is entirely false. With the advent of DNA analysis, we now know that a significant percentage of young in songbird nests are not genetically related to the male bird feeding them. To give just one example, about 14 per cent of house sparrow chicks across the board are born 'out of wedlock', so to speak. Similar percentages are recorded for other species, including swallows, while lower levels of cuckoldry are recorded for such birds as robins and blackbirds. You can say that, routinely among songbirds, a great deal of 'cheating' is carried out by both sexes during the period of egg formation.

Before having a look at why cuckoldry is rife, we should return to the matter of why it was overlooked for so long. Why did people assume that birds were generally 'faithful' to their partners when in reality, this wasn't so? There are probably four main explanations.

The first explanation is that birds are very difficult to recognise individually. In my own life as a birdwatcher, I have only been able to lock on to a few birds that are so distinctive as to stand out from the rest. A few have shown plumage abnormalities, such as white patches or missing feathers. However, since all birds of the same species and sex look similar to us, it isn't surprising that we don't catch individuals red-handed when they consort with birds that are not their mates.

Secondly, the period during which this intriguing behaviour occurs is very brief. A songbird female can be inseminated during her egg-laying period, which lasts from a week to a couple of weeks per clutch, depending on the species. Some females lay two or three clutches in a breeding season, but even then there is very little time to observe this going on. Furthermore, birds are usually coy about their extra-pair liaisons, because if they are caught in the act, this can have consequences – of which more later. And on the whole, the extra-pair copulations (EPCs) are simply that, rapid physical exchanges that are easy to miss.

A third reason is everything to do with coyness, but this time of an entirely human nature. It is easy to forget, in these modern times when we have learned to live with a certain scientific detachment, that until recently we placed our own values on animal behaviour. Our world view was so human centred that we could claim the extra-pair copulation of two birds as an infidelity, and therefore as a sin – and in the birds' world, while it is arguably the first, it isn't the second. From both a cultural and religious perspective people were unwilling to concede that the pair bond in birds could be anything but solid, and acknowledging such a thing as a widespread practice would have been nothing less than scandalous in, for example, Victorian society.

However, there is a fourth and most important reason why it has always been difficult to appreciate widespread 'infidelity' in birds' partnerships, and that is the behaviour of the birds themselves. Birds, particularly songbirds, look like a unit and, crucially, act as a unit. Both sexes have clearly defined roles within the breeding relationship and, apparently regardless of whether or not they have 'cheated', they play these roles.

With a few exceptions, the arrangement works out like this. At the very beginning of the breeding season, often as early as January or February (later in summer migrants), the male bird starts singing in order to lay claim to a territory; as already discussed in this book, females usually don't sing. The song spreads the males out and attracts the attention of females. Once a male has excited the interest of a female, it adds visual ritualised postures to its song display, and little by little the breeding unit is formed. The female starts nest building, sometimes, but not always aided by the male, which might also bring in deliveries of food to the female, serving her bill-to-bill (courtship-feeding). When the nest is complete the female lays the eggs and usually performs the lion's share of the incubation. The male often continues to bring in food deliveries at this time. Once the eggs hatch the female usually sits on the nestlings, a process called brooding. While she sits, the male brings in food for both female and young and will continue with this until the young have grown enough to need less brooding. Male and female will then, together, bring in food to nestlings and, later, fledglings.

For this to work, there must be close co-operation between the sexes. It is no good if the female tries to start a breeding attempt without the male having established the territory first because the neighbours would disrupt the process. It is no good the male failing to bring in food when the female requires extra rations to produce the eggs because many a female depends on this food. Each sex must be physiologically ready for every stage and both members of a pair must be physiologically in tune with one another – they must calibrate the other bird's behaviour. Many females, for example, are stimulated to ovulate just by hearing the male's song. All the tasks are fitted into a very short period of time. It is a highly complex set of behaviours which must fit together precisely.

With this in mind, you have a picture of a remarkable partnership between two individuals, with very clearly defined roles. A breeding attempt has absolutely no chance of succeeding unless the birds work together and know what to do and when to do it. Many young birds fail at their first breeding attempt, and many individuals of every population fail ever to reproduce at all. Whether it be singing, nest building, incubation or feeding the young, there are many pitfalls, and if either individual of a pair falls short, the whole breeding attempt is in jeopardy.

It is perhaps no wonder that everybody once assumed that a pair of birds was almost always a closed shop, both socially and genetically. It stands to reason; the only problem is that it isn't true.

Scientists now make a distinction between two types of monogamy: social monogamy and genetic monogamy. Genetic monogamy is self-explanatory. It is by no means unusual, either, and some songbirds such as wood warblers and jackdaws form what we could call 'faithful' pair bonds. Social monogamy, though, is the norm. In the majority of songbird breeding attempts the above arrangement applies, with each sex taking on a defined set of roles and tasks. The only difference is that, when the opportunity arises, both members of the pair routinely copulate with other members of the opposite sex. Remarkably, of those species for which DNA paternity studies have been carried out, 85 per cent show promiscuity within the social monogamy framework, and just 15 per cent are genetically monogamous. Look out into your garden at your beloved birds – they are all at it!

The consequences of extra-pair liaisons within the framework of a social working relationship are truly profound. At the same time, the potential benefits to an individual are equally obvious. As far as a male is concerned, the more females he can inseminate the better; the evolutionary urge to cast your genes abroad is a powerful one. It is better still if, at the same time as inseminating a female, another male can put in the effort to bring up your progeny – what's not to like?

The benefits for a female are more subtle, but no less important. The males in a population vary in terms of quality, in all sorts of ways, and not every female ends up with the most attractive, strongest or healthiest male as her social mate. That does not mean, however, that she cannot solicit the attention of the sexiest male on the block. As often as not she does just that. Really, the female is hedging her bets; the sexiest male is likely to father the highest-quality youngsters with the greatest chance of survival, and of attracting high-quality mates of their own in due course. If one or two of her brood come from the best stock, it will be to the female's advantage in the evolutionary sense.

The upshot of this arrangement is conflict between the sexes and individuals of the same sex – females in a group compete for the attention of the males, as we will see later. Social monogamy engenders a festering distrust between everybody and, during the females' fertile period, outbreaks of anarchy. So much for the genteel world of breeding songbirds!

As mentioned, the flashpoint in the breeding season occurs when the females are producing eggs. As a rule, eggs are not fertilised until the last moment; a female may hold eggs in her uterus for days before the deal on their paternity is closed. Most females lay an egg a day, just after dawn. In theory a female could be successfully inseminated any time during the preceding day, or even earlier, since it is now known that at least some birds can store sperm for ten or more days. What isn't in doubt is that the egg-laying time, the fertile period, is the window of opportunity for liaisons outside the pair bond.

As you might imagine, male individuals do not take kindly to the possibility of rivals inseminating their social mates and impinging upon their paternity. They don't just lie down and let their females wander at will. The danger of promiscuity has led to the development of mate-guarding behaviour, which is exactly what it sounds like. The male bird sticks close to his mate all day long, never intentionally allowing her out of his sight, and staying tight to her. He sometimes invades the female's personal space, which probably feels suffocating to her, but this is a high priority for the male.

Oddly enough, mate guarding can look to us like one of those apparently noble gestures that are associated with social monogamy. We observe a male great tit being extra attentive to his female and it looks like high-intensity courting behaviour when it is actually the opposite – high-intensity anti-courting behaviour. We see pigeons 'driving' their mates, almost stepping on their feet as they walk, and think of couples unable to keep their hands off each other. Yet mate guarding is a symptom of what I have called the dark heart of bird behaviour.

Nevertheless, the sheer desperation of mate-guarding males can lead to some amusing situations. The biologist Angela Turner has described how resourceful male swallows cope when finding themselves in the situation of accidentally losing sight of their mates during the guarding period. Breeding within a colony, where promiscuity is rife (swallows are notorious cheaters), a lost female is a serious threat to any male's paternity. So the male makes loud hawk-alarm calls, summoning everybody in the colony to attention. As an anti-predator response, the members drop what they are doing and fly up into the air above the breeding grounds, forming a closely compact flock of individuals – the idea is that they are close together and can all see the potential predator. When a hawk alarm is given, every swallow would be wise to respond or it could be putting itself at risk. Thus, despite in this case setting off an entirely false alarm, the male is likely to summon his wayward female out from hiding and into the flock's anti-predator melee, allowing him to re-find her, hopefully before she has copulated with the sexy male in the barn next door.

I have mentioned earlier, too, about the possible link between the dawn chorus and mate guarding. Male great tits typically sing loudly at dawn immediately outside the nest where their female has been sleeping, 'assuring' them that their mate is awake and attentive. At dusk, some keen or desperate male great tits have been observed following their mates back to the nest-hole and then remaining outside well into nightfall, when a female's illicit sojourn would be impractical.

Recent research by Tim Birkhead and his co-workers at the University of Sheffield has confirmed just how important this last-ditch mate guarding can be. They discovered that, in the case of the zebra finch, a common Australian dry-country bird that is easily kept in captivity, there is a phenomenon of 'last male precedence', meaning that the last male to copulate before an egg is laid tends to fertilise it. This occurs even if the female's social mate has repeatedly copulated with her throughout the day; if a male steals in at the last moment, the Johnny-come-lately often turns out to be the father.

If a male spots his female acting suspiciously or has reason to suspect her of cheating, he will, logically enough, mate with her on the spot if possible. Who knows – you might have witnessed this yourself when you saw some garden birds *in flagrante*, not realising that it was 'retaliatory copulation'. The very term demonstrates the nature of this insidious game of paternity that goes on in our songbirds, and indeed the whole subject is frequently called 'sperm competition'.

The effect of sperm competition is felt beyond the simple act of copulation. It spreads mistrust, so to speak, within the pair bond; each individual will be uncertain of their partner's commitment to the project of raising young with their shared genetic footprint. Obviously, the female lays the eggs, so she can be sure of their provenance. But for the male, his paternity is at stake, and any hint that the female has been involved in extra-pair sexual liaisons can affect his subsequent behaviour. He can respond to a cuckolding female with what might be called 'retaliatory laziness', which is essentially withdrawing effort from looking after the young. In the case of the reed bunting, a small sparrow-like bird of reedbeds and other wet areas, the males are able to monitor the number of illegitimate nestlings in their brood – nobody knows how. In this species there is a direct link: the more such nestlings, the smaller amount of food the males provide. Other species cannot tell and cannot recognise their nestlings.

But why should a male reed bunting reduce his efforts when his own young might suffer in addition to the foreign offspring? Unless he only brings food to his own young, which hasn't been proven, it seems that this ensures that there is a cost to cuckoldry for the female, as well as the male. The male's behaviour is making it less attractive for his mate to seek extra-pair copulations. She will suffer for it.

If ever there was a discovery that blows the lid off our birds' secrecy and unveiled their strife-filled existence, it must be this rash of everyday promiscuity. In a study of coal tits, for example, some researchers concluded: 'When analysing three or more broods of the same individual, there were no completely "faithful" females and also nearly no males which were not cuckolded at least once.'

An interesting bias that occurred in coal tits, which have unusually high levels of infidelity, is that the worst culprits are the experienced birds with age on their side. In the world of birds, as we have seen, a successful breeding attempt is a badge of honour, conferring great attractiveness on the fortunate individual. Younger females seek out experienced, older birds for extra-pair copulations.

Up until now, we have looked at your average songbird and found out a few general breeding rules – namely, that a high proportion of birds are 'unfaithful' to their partners and that many a brood of eggs incubated by the sitting female are likely to contain the genetic material of more than one male. However, there are a number of variations on this theme. In some species, for example, a male has a social relationship with more than one female, either simultaneously or in sequence, and each female is typically 'faithful' to him. In such cases there is one territory, held by the male, but two mates and the male will sometimes help with feeding both sets of young. These polygamous arrangements are quite distinct from extra-pair copulations, where there is no relationship between the sexes beyond the casual fling. Here the birds are 'formally' partnered; it's just that there are two or more of one sex (occasionally both sexes, see page 76). Sometimes a female acquires several male mates (polyandry), but more often a male has two or more female partners (polygyny). On many occasions the relationships are quite open, and the females intentionally share the male's genetic charms. Polygamy is something of a biological can of worms, with complex interrelationships.

Perhaps surprisingly, a good example of a polygamous bird is a very familiar one, the wren. In fact, this species is the commonest bird in Britain, with an estimated 8.5 million breeding pairs. Except that not all wrens nest as pairs – around half are monogamous, but the rest of the males either don't pair at all or pair up with two, three or occasionally four females. The females mated to a male use his territory, and finish off the preliminary nests made by the incumbent; normally they don't all breed at exactly the same time, but one female follows the other, with one at the nestling feeding stage and the other, say, at the egg-laying stage. The relationship between the sexes isn't close, but the male will often make a contribution to bringing food for the young. The amount of effort a male makes partly depends on how many females he is housing within his territory. There is little or no doubt that each female is aware that others are sharing the same mate, but the arrangements have a somewhat permissive feel. If there is a well of aggression between mate-sharing females, it remains below the surface.

Another secretive bird with similar arrangements to the wren, but which is far less often monogamous, is a songbird that only entered the British consciousness when it first arrived in 1961 as a new species for the country – the Cetti's warbler. Formerly a Mediterranean species, this secretive bird crept ever northwards through France for a few decades until making landfall here, and its population has thrived since the first breeding in 1973, in Kent. There are now at least 1,900 territories in the UK but, once again, the population cannot be quoted in pairs. In one study in Weymouth, Dorset, nearly 60 per cent of males were polygamous, with half of these bigamous and, impressively, a third managing to hold on to three mates. The odd exceptional individual was paired with four females. As in the case of the wren, not all the females would breed at the same time, rather breeding would be staggered through the summer. Males often help with feeding their young, especially second broods.

One particular feature of the Cetti's warbler's polygamy is especially interesting and perhaps surprising – it can be a repeatable arrangement. Not only do *ménages à trois* remain in place from one brood to another in the same year, but they can reappear the following breeding season, suggesting that the females are perfectly content with what has gone before. Some female Cetti's warblers kept pairing up with the same male for up to five years, regardless of whether the relationship was exclusive or not.

You could conclude that there wasn't much pain involved here, with a thoroughly modern arrangement favouring everybody. That, though, would be to forget about the unpaired males of the neighbourhood. The problem, if you could call it that, with any kind of polygamy is that somebody could be left out.

Quite a few males have missed out in some populations of another small brown bird with polygamous inclinations, the corn bunting. This plump creature is the absolute quintessence of drab, brown, streaky, featureless anonymity. On many occasions I have pointed them out to people as they sit on wires or weeds within fields, and have felt almost embarrassed to mention them, even apologetic. You cannot tell a male from a female and both are at best sparrow-like. Even the courtship display of a corn bunting is a derisory affair: the bird flies from one perch to another with its legs dangling down – and that's it.

The brown streaky bird of the brown, streaky weedy fields does, however, have its lively side. In common with other polygamous birds, only a percentage of the population actually partakes in multiples, while many of their colleagues are solidly monogamous. What is remarkable is that, on occasion, a male will take on an obscene number of partners. One individual in a Sussex field actually paired with eighteen females in a single season, including six at once – I think that is a British record for polygamy. It is unrecorded whether he survived for long! One thing is for sure, though: he made a lot of rivals unhappy.

Besides, it would be wrong to assume that polygamy is always a serene understanding between paired birds. In the case of the pied flycatcher, there is no understanding of any kind. If anything, it is one bird abusing another.

Pied flycatchers are attractive migratory birds that are found mainly in the western and northern half of Britain, preferring sessile oak woods. They nest in holes in these trees or, if available, in nest boxes – indeed pied flycatchers actually prefer the boxes to natural tree holes, making them easy to study. The *modus operandum* of a male pied flycatcher is to attract a female and, once the pair bond is established, to attract another – this is successive polygyny. This bird is very sneaky about it, however, owing to one very unusual quirk – he seeks to form two territories, not one.

Let's assume a male pied flycatcher arrives in good time on the breeding grounds, in mid-April. In common with most of the other small birds of the woodland, he will sing to claim his first territory, with a short, slurred ditty delivered with spirited effort but little musical appreciation. Nevertheless, the first females arrive a couple of weeks later and seek out males, which usually perch at the entrance to the nest box or just below it. Things take their course and birds pair up. While the male defends the territory assiduously, the female builds the nest. They copulate and the female produces a clutch of eggs. In most small birds incubation takes about a fortnight and for the male pied flycatcher this is a time of opportunity. With a female safely ensconced in his territory and on eggs, a significant proportion of the males then endeavours to attract a second female during the inevitable down time. The way they do it is mischievous, indeed downright deceptive.

The pied flycatchers do something that very few other songbirds have been found to do, and that's to procure a second territory. They move a safe distance away from their primary female, which can be anything from a few tens of metres to, remarkably, 2km (1.24 miles). Here they find another potential nest-hole and sing to attract a mate as if nothing has happened. Certain females duly oblige, and the two-timing pied flycatcher mates with the secondary female, which lays a clutch of eggs. The moment the clutch is complete coincides roughly with the time that the primary female's eggs begin to hatch, ushering in hard work for all concerned. At this point the male abandons the secondary female and concentrates entirely on feeding his first (primary) brood. The secondary brood, when it hatches, doesn't usually fare well, but any surviving chicks are, of course, a bonus for the male.

The system goes by the tongue-twisting name of poly-territorial polygyny, and in common with many of these more unusual arrangements tends to coexist with monogamy and a degree of promiscuity. Although there is a clear benefit to the male, the secondary female would appear to suffer, but she might be partly hedging her bets, especially if she arrived late on the breeding scene. A crack at a breeding attempt, even with a two-timing male, could well turn out better than not attempting to reproduce at all.

Nevertheless, the pied flycatcher's antics do show how deceptive birds can be. Why would any bird hold two territories hundreds of metres apart if it isn't to deceive a mate – otherwise, why not just keep them both in the same territory?

For sheer soap opera, the complicated sex-life of a very common garden bird beats every other. The dunnock, also widely known as the hedge sparrow, is one of our top twenty most familiar neighbours and everything about its appearance speaks of modesty and restraint. Until recently it was famous for being unobtrusive, for taking a back seat and, indeed, if you can be well known for being overlooked, then the dunnock fitted into that category. Plain brown and streaky, with a pleasant grey wash to the head and upper breast, the dunnock is a retiring, ground-feeding, sparrow-like bird that rather shuffles along the ground and frequently flicks its wings and tail. The eyes and the legs are reddish. In contrast to a sparrow, the bill is very thin, especially at the tip, and it feeds on very small invertebrates and even smaller seeds.

At the risk of sounding coarse, if you were able to look under the dunnock's plumage in the breeding season and see the male's huge testes and the female's large extensible cloaca (the reproductive opening) you might just have suspected that something was unusual. But until the 1950s, nobody did. Indeed, in his rightly lauded book *A History of British Birds*, written 1851–57, the Reverend Frederick Morris sang the praises of seemingly wholesome dunnocks:

> *Unobtrusive, quiet and retiring, without being shy, humble and homely in its deportment and habits, sober and unpretending in its dress, while still neat and graceful, the dunnock exhibits a pattern which many of a higher grade might imitate, with advantage to themselves and benefit for others through an improved example.*

It was written at a time when birds were frequently held up as a moral example to humankind. But, oh dear, in hindsight the dunnock was the wrong bird to choose.

Again, with hindsight, it is surprising that dunnocks were never rumbled, because it is not particularly difficult to see them behaving oddly. Only recently I was taking a group of executives on a day out birding in the New Forest, so they could relax and immerse themselves in nature. At lunch, we went to a delightful pub, the Royal Oak in Fritham. Here, in broad daylight, in front of everybody, on the pub lawn, two dunnocks met in the grass, with much flicking of wings and tails. One bird then suddenly stood still, crouched, shivered its wings and lifted up its tail unsubtly to the bird behind. The second bird then proceeded to peck hard in the direction of its cloaca. This was so obvious that the executives watched in rapt attention – and these were people who weren't used to birding. How can the observers of old have not seen this? And if they had, surely they would not be recommending acting like dunnocks to their parishioners?

You are probably wondering right now what shenanigans dunnocks get up to. Well, it's complicated. The short answer is, almost anything of a heterosexual persuasion. The longer answer is that they will practise most forms of pair bonding, including monogamy and both forms of polygamy, polygyny (one male, two or more females) and polyandry (one female, several males). A dunnock can be involved in all these arrangements in a single breeding season. That's because they sometimes mix polygyny and polyandry in the same territory. And, on top of all this, both sexes regularly indulge in extra-pair copulations completely outside the monogamous or polygamous pair bonds in which they are already involved, adding promiscuity to all the polys.

The true delight of the dunnock's story lies in how its behaviour is tied in with its ecological requirements, and it is also a classic case of conflict between male and female and between males. So, are you sitting comfortably? You can read the full story here.

It all starts at the beginning of the breeding season, when both sexes sing to defend territories. It is very unusual for females of any British birds to sing for territory at all, but dunnocks are an exception. In most of the gardens, hedgerows and woodland edges of Britain, the pool of males tends to be squeezed in comparison to the females, so the more densely packed males have smaller territories than the females.

This is unfortunate because once a male eyes up a female, it is incumbent on him to take over her entire territory and defend it. Some males manage to do this, because they are strong enough, sing well enough and are solid citizens. The stage is set for monogamy. Being typical small birds, the arrangement is only social monogamy, and both birds will take the opportunity to copulate outside their pair bond. Nevertheless, it is genuine social monogamy, with all that this entails, the pair dividing duties. In addition to taking on the defence of the realm, the male will devote much effort to feeding the young, both as nestlings and, later, as fledglings.

Occasionally, but this is rare in dunnock society, you do get exceptional individuals, males that are so fit in every way that they could take on more than one territory, and the females that go with it. These are equivalent to men in human society that can be company directors, give plenty of dad time to their children, remember to buy surprise gifts for their wife and their mistress, and become British unicycling champion, or whatever, in what spare time remains. And never lose at Scrabble, either. Dunnocks of this type have been recorded taking over and successfully defending two neighbouring female territories, and thus becoming polygynous. One would imagine that the females paired to these males would be uninterested in acquiring genetic material from another source, although it probably doesn't work the other way round.

It is routine, however, for males to find it difficult to sustain protecting the borders of a mate's larger territory. In such circumstances it would seem that they have no choice but to allow the help of another, unpaired male. Bearing in mind that this is equivalent to allowing a young, randy lodger to share your house with you and your young, randy partner, you can imagine that males do this somewhat reluctantly. Their reluctance, indeed, can be measured by the often violent fights that take place between the sitting male and the prospective tenant. The original male, though, is caught between a rock and a hard place; without the extra help he might lose the territory and reduce his breeding success to nil, but the presence of another male is going to be an ever-present threat to his paternity.

The female, meanwhile, is in a win-win situation. Having two males sharing the defence of the territory takes care of that problem. And having two males in close proximity introduces an obvious extra benefit. It isn't that the female is keen on copulating with anybody and everybody; that isn't in her interest, and the chances are that her original partner, which we will now call the alpha male, will be superior to the incomer anyway. The female is being much more practical. If she allows the beta male to copulate with her on a regular basis, he is then likely to put effort into feeding her progeny alongside the alpha male. If both males think they are feeding chicks that they have fathered, they will evidently put in the effort. Being fed by two males gives them an enhanced chance of survival.

While the female is secretly delighted (there is doubtless a more scientific way to put this, but you get the idea) at the presence of the second male, the alpha male is only too aware of the threat posed to his paternity. The tension between the males crackles; they frequently fight, and it is by no means unknown for one male to kill another. Every soap opera needs a murder. It is also quite easy for you and I to see these disputes boiling over, because they are marked with a common and distinctive display. The protagonists face each other and flick their wings, or more obviously, they wave their wings slowly. Sometimes they wave both wings together, and sometimes alternately. Either way, wing waving is not friendly.

Within the *ménage à trois*, both males have a strategy. The alpha male is a fanatical guard of his mate (see pages 63–65). He will do his best to keep the female in sight and to interrupt any high jinks. However, bearing in mind that the female is likely to be fertile for ten days, the alpha male must also find time to feed and to attend to any other essential duties. He might simply lose concentration, whatever the avian equivalent of losing concentration might be. Any time that his guard drops, the beta male could steal in and copulate with the female.

The beta male's strategy is simpler and easier. He just needs to be persistent. Opportunities will present themselves, in the ways described above. The beta male's main problem is the extreme aggression of the alpha male. He will be well aware, however, that his paternity will be zero if he does become cowed.

The beta male's lot is greatly eased by the attitude of the female. As mentioned above, it is in her chicks' best interest to be fed by two males, so she acts in co-operation with the beta male. In the equivalent situation within human society, she will keep texting the beta male to invite him behind the bike sheds (in the University Botanic Gardens in Cambridge where much of the dunnock study was done, it was apparently sometimes literally behind the bike sheds). There is an elaborate game being played out, equivalent to a comedy farce, with the couple attempting to evade the resident male at every opportunity, ducking hither and thither, hormones ablaze, shuffling into thick vegetation, being discovered, acting warily. To the credit of the alpha male, the beta male sometimes fails in his attempt to inseminate the female.

You won't be at all surprised to hear that the copulation display of dunnocks is elaborate and nuanced. This was what my executives' group watched at lunch in the New Forest. The female solicits and lifts up her tail to expose the cloaca and, over the course of about a minute, the male pecks it about thirty times, until it is distended. The cloaca begins to make a pumping action and eventually discharges a drop of sperm, which could be from a previous illicit copulation. Satisfied that his paternity is safe for the moment, the male then mates with the female. You might think that repeated, extended copulation might now be the order of the day, after all, that might or might not have gone before, but the physical cloacal contact lasts for less than a quarter of a second, and is easy to miss to the human eye. The birds go their separate ways, nobody sure of anybody.

You might think I'm joking if I say that the arrangements laid out so far are the simpler options, but the web can become yet more tangled, with other characters entering the cast. On rare occasions, when two males are involved in defending a female territory, they find that together they can defend not one, but two adjoining female territories. On these occasions it becomes, so to speak, a *ménage à quatre*, with two males and two females occupying the confined space. You might think that, in such a case, the two males and two females within the arrangement might simply revert to two monogamous pairs, but this seems not to happen. Instead the alpha male and the beta male bicker over access to both females. A quite different conflict also now emerges between the two females. After all, they are competing for the males' attentiveness to their chicks. If each female manages to mate with both males, then both females will be assured some fatherly help with providing food for their respective chicks. However, both females would much prefer to have exclusive help from one or both males. This sets up a hostility as virulent as the earlier male-to-male conflicts. Each female seeks to prevent the other copulating at all and, if she fails, she might instead attempt to interrupt her rival's breeding attempt, even going so far as to cause it to be abandoned completely. Remarkably, the males actually try to break up the fights, with varying degrees of success. This arrangement, in which a male is paired with two females, who are each effectively paired with him and another male, is known as polygynandry. To say it isn't the ideal for the parties concerned is probably stating the obvious.

The researchers discovered that polygynandrous arrangements sometimes encompassed yet more individuals, up to four males or females, evidently all attempting to copulate with each other. It is important to remember that, in a sense, these are formalised pair bonds. If three male dunnocks are sharing a territory, all will sing to defend the boundaries and, provided they have managed to copulate with at least one female, they will help to feed whichever broods they have part-fertilised. This is not a case of uncontrolled promiscuity, in which a male takes no part at all in a breeding attempt beyond providing genetic material. Hard though it is to believe, both sexes also sometimes invade neighbouring territories to indulge in extra-pair promiscuous copulations. Or should that be extra-polygynandrous copulations?

Since I have persisted in calling the dunnock's arrangements a soap opera, scientifically dodgy though that undoubtedly is, I am going to make a brief foray into a couple of modern-day conundrums that probably pass the dunnock by. No soap opera is complete without a gay character, or often a transgender character, but, on the whole, this isn't relevant to birds. They cannot change the sex of their physical bodies in the way that, for example, fish can. Individuals are occasionally born hermaphrodite but in the wild they are anomalies. As for homosexuality, this occurs quite widely in the animal kingdom, as does incest, and both occur in bird populations. If a bird is entirely homosexual in orientation, in direct reproductive terms it will obviously be passed by.

Another classic soap-opera issue would be the effect of split parenthood. In human society, divorce during parenting can cause many difficulties, but humans are exceptional for the depth of their long-term parenting. It would seem that all that matters for birds is exactly what the female dunnocks are concerned about – the amount of food their progeny receives in their early life. It doesn't matter whom the provider actually is.

The unusual behaviour of dunnocks does not occur in a vacuum. It is actually related to food supply. You will remember at the start that dunnocks pair up when a male attempts to take over the defence of a female's territory, and his ability to defend it is related to its size – well, the size is also related to how much food is available within. When the researchers in Cambridge provided extra food to dunnocks experimentally, the females adjusted their territories to be smaller. It seems that the females know how much food they need. So if, for example, the territory is particularly rich, the area could be as small as 0.23 hectares (2,750 square yards). The researchers found that this was an easy size for just one male to defend – indeed, a good male could defend a couple of territories this size on his own, thus pairing up with two females and becoming polygynous. On the other hand, the scientists discovered that if the female's original territory covered as much as 0.55 hectares (6,576 square yards) on average, because it wasn't as rich, then it would take a couple of males to defend it, making polyandry far more likely.

If you think about it, the system is brilliant. A female defends a relatively poor territory, where food is somewhat difficult to find. So she makes her borders wide, difficult for a single male to defend, thereby necessitating two males to share the defence. Having attracted two males in, she plays them along and – hey presto! – she has acquired two labourers to help feed the young. On the other hand, if she has a territory with very rich food resources, she can maintain tighter borders because she only requires a single male. This is how the female should be able to ensure her young are fed well. Of course, the female is not intentionally 'scheming' in the human sense of the world. She is merely reacting to the density of food supplies in her neighbourhood, and the rest follows.

As we know, though, by defending a large territory and compelling two males to defend it for her, the female causes intense and often violent competition. The same thing happens when females have to share mates in polygyny and polygynandry. Their motivations are different, though. As far as males are concerned, they want to maximise their paternity, while females want to maximise the food brought to their young. In situations where a female does have to share male provision, the young are more likely to starve and, if they do fledge, start life in a poorer condition than those served by two males, or by a male unpaired to another female.

The breeding system is the result of this conflict of interest. In a remarkable long-term dunnock study, a team was able to measure the breeding success of individuals in each arrangement. Nick Davies and his colleagues at Cambridge University found that, in a monogamous system, both male and female, on average, would raise five chicks. However, it would be highly advantageous to a male to be in a relationship with two females. On average, such a super male would raise 7.6 chicks. His two females, however, would lose out, raising just 3.8 chicks each. On the other hand, for a female, a polyandrous relationship is the best option. A female paired to two males would generally produce 6.7 chicks, well above the monogamous rate of five. However, both of her males in this system lose out. Alpha males in a polyandrous system might expect to bring 3.7 chicks into the world, with their rivals, the beta males, managing just three chicks each. Alpha males come out fairly satisfied in a polygynandrous arrangement with two males and two females; they expect to bring up five chicks, while their beta male rival only brings up a pathetic 2.2 chicks. The two females successfully bring up 3.6 chicks each, not a disaster, but nothing like as good as going it alone with one male.

It is hopefully fairly obvious from the above statistics that monogamy is a satisfactory arrangement for both sexes where the food supply is rich enough to sustain it. However, it is highly tempting for male dunnocks to aim for polygyny and for females to aim for polyandry. What seems to happen is that these aims are sometimes dashed, leading to polygynandry, which is very much the compromise position for everybody.

The extraordinary complexity of these formalised arrangements is apparently matched by their fluidity. A bird might settle into a monogamous relationship at the start of the season, only to end up in a polygynandrous compromise for its second brood, and things can change from one year to the next.

Sexual tension, murder, infidelity – such is the widespread tendency for songbirds to adhere to social monogamy or more tangled relationships, one can almost sense the researchers floundering when they discover a bird that actually exhibits genetic monogamy, equivalent in human circles to a marriage free of affairs. Such a bird is the jackdaw, a common member of the crow family that often nests in chimney pots in towns and villages. Say the authors in a 1983 paper, Henderson, Hart and Burke: 'The parentage analysis revealed no cases of extra-pair fertilisation.' This is presented as a major surprise, especially because 'nest-sites are a limited resource, non-breeding adults are also common around colonies and males are frequently separated from their incubating females during the fertile egg-laying period… opportunities arise for extra-pair copulation.' They go on to ruminate on what it is that 'constrains' the species to true monogamy.

And 'constrained' they are. Jackdaws are exceptionally faithful to their partners, even among genetically monogamous birds. Once a male and female have been paired up for more than six months, they virtually never separate. There is one instance which often splits up monogamous couples in the bird world, and that is consecutive breeding failures. Even among albatrosses, which are extremely dependent on their mates, breeding failure can be cited in the divorce papers. Yet in jackdaws, even this doesn't happen. It is genuinely a case of 'til death do us part'.

The question is, why? It would appear to be related to the fact that it is very difficult to find enough food to nurture jackdaw chicks. They are fed on caterpillars, flies and beetles, and it appears to take extreme hard graft from both sexes working together to acquire adequate nourishment. From a female's perspective, it is absolutely essential that the male does his bit, and the only way to compel him to do so is to 'assure' him of his paternity. We have already learned that there is a negative link between the male's parental efforts and the amount of extra-pair chicks in the brood. In the case of the jackdaw, extra-pair liaisons are avoided by society as a whole, meaning that males can put in their parental feeding efforts without compromise.

The occasional need for genetic monogamy throws up some unusual bedfellows – although that's probably not the phrase to use! In addition to jackdaws, only willow tits seem to show high levels of faithfulness. Among birds other than songbirds, albatrosses, fulmars and great northern divers are similarly averse to sex outside the pair bond.

However, as we know, these species are the exceptions. The ever growing body of research now shows that very few birds would keep to their wedding vows if ever they made them. It is curious that research on birds has revealed bird society reflecting the zeitgeist, a permissive human society uncovering a permissive bird society.

Yet songbirds, as we know, live the briefest of lives, with the slenderest of chances to pass on their genes. Outside of pure survival, reproduction is songbirds' overwhelming motivation. The litany of behaviours described in this chapter doesn't sit easily with us, perhaps, but we shouldn't be surprised. The most basic of urges doesn't allow room for sensibilities.

CHAPTER FOUR

Competitive Exclusion

Every day throughout Britain, but especially in the winter months, birds visit feeding stations in gardens for food. Human beings, who have put the food out, think this is marvellous. It is a quick and easy way to connect with the wild. Birds come from outside the garden and partake of the easy pickings, and human beings would like to think that the birds were grateful. A succession of individuals comes along on any given day and feeds and maybe visits the water put out, too. In most gardens, several species come: almost always blue and great tits, and coal tits as a bonus. Blackbirds, dunnocks and often chaffinches feed on platforms or on the ground. Exotic species such as siskins are winter visitors, and some gardens are blessed by rose-ringed parakeets. A really good garden, with many feeding stations, will attract a significant number of birds.

People invariably put out food for good reasons. Yes, they enjoy seeing the comings and goings of wild creatures for their own pleasure, but mainly they want to help. And there is no doubt that they do. There have been a number of studies recently that show that garden bird feeding helps individuals to survive the winter. Indeed, some survive in a garden context when, unaided by extra provisions, they would undoubtedly die. This has led to the question of whether we are enabling weaker birds to propagate substandard genes, in a curious case of the survival of the not so fit. But whether this is the case or not, people are going to carry on feeding birds and enjoying it without a shred of guilt.

But this book is about the harsh realities of a bird's life, and one inevitable consequence of feeding birds cannot be shirked: you are going to induce conflict. You will intensify conflicts that are already there, and you will cause intense competition that would never otherwise have happened. The feeding station isn't quite a war zone, but it isn't a supermarket on a quiet day, either. Cast your mind back to those pictures you've seen of department stores offering irresistible, once-in-a-lifetime bargains on sales days. There can be a scrum of people behaving with aggression, and real fights breaking out – and this isn't even a scrap over a life-and-death resource. It's to obtain a widescreen TV so that the winners can slob in front of atrocious programmes. Well, the food on your bird table *is* a life-and-death resource. At the end of winter, and beforehand if conditions become freezing, finding your bird feeders is a lottery win for any bird. And funnily enough, as seems to be the case with human lottery winners, conflict is magnified.

I will quickly mention here that another consequence of feeding birds is predation, by sparrowhawks and cats, for instance. By putting up a bird feeder, you are inadvertently creating a fast-food outlet for predators. I will discuss this in chapter 5.

Competition for resources among birds is bad enough, even without your kind-hearted provision. Consider a blackbird feeding on your lawn. It has a fixed way of feeding. It will hop or run along the lawn and suddenly come to a stop, remaining still for a few minutes, perhaps cocking its head quizzically sideways. After a while, it will break into another smooth run, and this time when it stops it might peck at the grass or lunge towards a worm, tugging it out with ruthless expertise. Slightly better fed, it returns to a standing position before trotting off once more, making short movements like a rounders player making it to the next base.

This ground-feeding style requires concentration on the part of the blackbird, but it also requires peace and quiet from its environment. This is partly in the form of low ambient noise because blackbirds typically detect the movements of worms in the soil by ear. Vitally, though, the blackbird wants the lawn to itself. Disturbance is a ground-feeding nightmare. The presence of another bird on the same lawn is not ideal.

Some scientists measured the feeding rate of a blackbird on a lawn. When a second blackbird arrived nearby, the feeding rate of the first bird dropped by as much as 43 per cent, and each time by more than half. Apart from feeding rate, though, there were no symptoms of conflict. The blackbirds concerned didn't fight or chase. They didn't acknowledge one another at all. The tensions remained invisible, but the effects were stunning.

And that is a conflict that doesn't even express itself. How about the battles on the bird feeder? You never need to watch your feeding stations for long before one bird flies towards another one and displaces it. It happens quickly and you easily miss the crux of the action, but if you have ever seen the films of birds at a feeder in slow motion, you can only then appreciate the potential violence of these so-called 'supplanting attacks'. They are all bills and feet. It's often big bird against little bird, and the big birds usually win. To make matters worse, the winners sometimes monopolise the perches for quite some time, taking in their fill while other birds wait on the periphery.

The greenfinch is a good example of the sort of bird that will do this. Greenfinches have something of a look of the gangster about them, with their big heads and slight frowning expression. Their bills are heavy and sharp, useful for cracking open sunflower seeds and managing rose hips, and the birds also seem to have an aggressive streak. They eat a wide variety of seeds and those bird feeders with perches provided make them comfortable. They give way to very little – a sparrow sometimes, and always to a nuthatch or woodpecker.

Much is made of the make-up of species at feeding stations, and it causes garden bird enthusiasts a degree of irritation. I have to admit it causes me a degree of hilarity. I love the sheer sense of outrage people express when the 'wrong' birds land on their feeders. Where bird tables are concerned, it is usually pigeons.

I frequently go around the country giving talks on garden birds, and at the end of the evening there is often a question-and-answer session that goes something like this:

'Can I ask you what to do with my pigeons?' asks the earnest gardener.

'What seems to be the problem?'

'Well, they keep coming to the bird table and not getting off.'

'Why is this a problem?' I tend to ask at this point, impishly.

'The food isn't for them. It's for the blue tits and other smaller birds.'

'Why, do you prefer blue tits to visit, then?' I am pushing them deliberately.

'Oh yes, I don't like pigeons. They are fat and greedy.'

There is always a murmur of agreement when somebody disses pigeons or doves. Somebody else might call out 'They trample my dahlias,' or some similar crime.

'Have you told your pigeons they aren't welcome?' I go on.

Such an exchange is light-hearted, but there is a serious point to be made here. If we do feed birds in our garden, should we be annoyed when they actually come, even when they are not our favourite species? And why is it that people disregard pigeons and doves so much? They are great entertainers in the garden, with evocative songs and eye-catching displays, and they are certainly willing visitors. Do we dislike them simply because they are plump and ungainly, as opposed to the sweet little colourful tits and finches? If we truly dislike pigeons visiting our feeding stations, is this because we wish to be in control of nature and mould it to our taste?

I don't make any case for rose-ringed parakeets, which many birders prefer to call ring-necked parakeets (but rose-ringed is the 'right' name according to the ornithological authorities). These are birds of the Sahel region of Africa, and they also live in India and surrounding countries. They survive here in the winter partly because of handouts. I remember seeing my first one in my south-west London garden in the early 1970s, not long after there were a series of introductions around the area of Heathrow. They were imported here as caged birds but some escaped into the wild and others were deliberately set free, and they have gained a foothold in several parts of the UK. The first 'wild' breeding was in Kent in 1971, but their real stronghold is the Greater London area. Further north, they are flourishing in Birmingham, Manchester, Liverpool and Sheffield, but are widely scattered elsewhere. They monopolise bird feeders and make a fearful screech. With at least 30,000 present in the late summer, these birds are likely to increase in both numbers and range in the near future.

Rose-ringed parakeets are a localised issue, but where they do occur they join in with the ranked species list, with the dominant species at the top and the subordinates below them. I have seen great spotted woodpeckers remove parakeets, but nobody gets past those two unless a jackdaw swoops in to have a go at a hanging feeder. In most gardens, however, the most obvious species-hierarchies are among the different tit species, all of which are subordinate to nuthatches, with their very long, sharp bills. Strangely the blackcap, a bird with a similar-size body to a great tit, seems to drive off all its similar-sized rivals by dint of its aggressive temperament.

Have you ever noticed, however, that none of these birds actually fight? Or at least, they very seldom do. Only today I was watching a feeder that was receiving visits from nuthatches, greenfinches and tits. Every time the nuthatch came down, the other species retreated, although the greenfinches would sometimes return before the sharp-billed heavyweight had left. But these were invariably bloodless encounters. Birds knew their place and only the promise of threat was required to maintain order. It was all shadow boxing, yet it was also decisive.

The reason actual conflict is avoided is that physical combat is highly costly, potentially to both protagonists. Fighting birds waste a great deal of energy and put themselves at great risk by reducing their vigilance. Distracted birds are frequently attacked by predators. And fighting incurs a risk of injury, by bill or feet. It isn't worth it. So hierarchies are maintained and are lasting.

Where tits are concerned, most gardens only have two or three species. Great tits are usually the dominant species, blue tits have roughly equal status, despite their smaller size, and the coal tit, which is really a specialist of coniferous woodland, is always ranked lowest of the three. If you are fortunate enough to have a marsh tit in your garden (those smart, plain brown tits with glossy-black caps and a 'Hitler' moustache), they slot in below blue and great tits and above coal tits.

Coal tits are smaller and lighter than the other tits (apart from the long-tailed tit, which is actually not a true tit at all). They have evolved to nestle into the needle-laden branches of conifers, where they acquire small insects that are out of reach of competitors. However, away from their core habitat they can struggle. If a blue or great tit sees a coal tit that has just unearthed a good meal, it will routinely try a supplanting attack. It just can't resist it. In common with us all, a blue tit is lazy and will fly at the chance of free and easy food. It knows a coal tit won't fight back and will yield the food item if challenged.

If you have coal tits coming to your feeders, you will probably have noticed that, while tits as a whole tend to bustle back and forth, coal tits tend to make their visits faster than other species. They steal in, often to the very bottom of the feeder, get the seed or nut as fast as they can, and then fly away out of sight. If we had to run the gauntlet of a gang of youths outside a pizza restaurant, we would probably hurry in and out with similar precipitous haste. The coal tit gets bullied so often that it has evolved a coping strategy for being constantly subjected to supplanting attacks. It is rather sad, but this bird seems to live in a cruel, cold regime of persecution from other, bigger birds.

The coal tit does, however, have a behaviour that could partly mitigate the effects of bullying. When it finds a good food source, such as a bird feeder, it will spirit away much of what it collects, storing nuts and seeds in the branches or needles of the nearest conifer or another appropriate tree, or on the ground. In contrast to acorns stored by jays, which are a long-term food larder, most coal tit rations seem to be consumed within a day or two. Is it possible, then, that the coal tit hides away its food before it can be spotted, then consumes it at leisure? There may be other reasons for storing food, including the obvious point of having food stashed away as insurance against bad weather, but the storage habit could also help it to keep away from bullies.

Of course, between-species hierarchies are only one type of 'pecking order' into which birds informally organise themselves. Perhaps of greater importance to most individuals' survival is how they fit in with other members of their own species. No bird flocks are ever full of equal individuals, and they all have hierarchies. In the case of tits, birds often spend much of the off-season in groups, and the individuals within these get to know each other well. As we've seen, birds recognise each other individually just as easily as we humans recognise our friends and colleagues; we shouldn't be surprised at this, just because we cannot easily tell individual birds apart. Humans in close communities quickly latch on to their neighbours' strengths and failings, and in bird society these are equally obvious to all.

Nobody is exactly sure how bird hierarchies actually form. Sometimes one bird will be self-evidently stronger than another – in the case of great tits, for example, the dominant males tend to have broader breast-stripes. However, in order for a hierarchy to form between closely matched individuals, there must presumably be history between them, a genuine fight perhaps, or a series of supplanting attacks. Once the status of each bird is fixed, it doesn't normally change much, if at all, in a season. The birds don't want to waste energy, time and, especially vigilance, on constant bickering day to day; both birds would lose out. Instead, it seems that most hierarchies are maintained without bloodshed.

Siskins, though, have a most unusual way of smoothing over relationships. These small finches live in flocks in the winter and, even when they are breeding, are highly sociable. They routinely visit bird feeders as a flock and, here, particularly towards the beginning of the breeding season, you might just witness something odd happening at the niger seed. Most people are familiar with a practice known as 'courtship feeding', in which a male brings food to a female and transfers it bill to bill. Not only does this help the female to be well fed, it also strengthens the bond between the sexes so, as its name suggests, it is part of courtship. In siskin flocks, however, you can occasionally see a male giving food to another male in the same way. This is not courtship, though, it is appeasement. The male proffering food is always subordinate, and the receiving bird is dominant. It seems to be a way in which the subordinate makes peace with the superior bird, presumably assuring that it won't be supplanted unnecessarily.

As mentioned above, a pecking order seems to click into place relatively quickly for most individuals. But how do the birds judge each other, particularly when birds are well matched? And how about males and females?

If the American black-capped chickadee is anything to go by, there are many ways in which a bird's rank can be acquired. This small bird, a great favourite of garden watchers and researchers on the other side of the pond, has been closely studied to within an inch of its life and vies with our own great tit as the best known wild bird in the world. Chickadees are judged on a multitude of variable characteristics, including how old they are, what sex they are, whether they are resident in an area or just visiting, as well as on their body condition, their plumage brightness, their singing ability and how early they begin the dawn chorus. All birds scoring well in this department are likely to be highly ranked individuals.

If we take a closer look at some of these chickadee characteristics, we can draw some general conclusions of what we might call universal laws of dominance. The greatest of these, seemingly important in most or all hierarchies that have been studied, is age, at least among males (and probably also females). Age takes precedence within sex, so older males are always dominant over younger males. It isn't just a case of age before beauty; it's a case of age being beauty. It doesn't matter how utterly gorgeous a bird is, it will always be usurped by an older individual. Age denotes current value; beauty is merely promising. Of course, in bird society, age equates to experience and survivability, precious characteristics indeed. In a society where very few individuals reach more than one or two years old, to have bred successfully in the past is to reach the top of the tree.

In the case of sex, males dominate females in all or most of the hierarchies studied, presumably for the simple reason that most are larger. In those few birds in which females are larger than males, such as raptors, females dominate males.

Another evidently important characteristic is residency, particularly when birds are of the same sex. In a number of winter flocks that have been studied, an element of society is resident and another element wanders between sites, moving through and not staying long. At a feeding station, transients are usually subordinate to residents. They also suffer far more aggression than do birds that are familiar with flock mates and surroundings.

It might not have immediately dawned on you reading this, but there is much potential fun to be had armed with this information. You can work out the characters in your own garden bird maelstrom, or at least, take a guess at them. If one bird supplants another of the same sex, you can make the assumption that it's older. If you see an individual flying aggressively towards another and flushing it off its perch, where the sexes look the same, you can assume it is a male supplanting a female. You can guess that a resident is supplanting a transient. Apart from guessing this for your own interest, you can tease your own neighbours and friends. You can pretend you know exactly what's going on. So you might say: 'Oh, that great tit that scared off the other, it's an older male.' Or say 'It's scaring it off because it's a stranger', and so on. You will probably be right!

You might be surprised to hear this, but personality, of all things, plays a part in a bird's social life, as it does in ours, and probably contributes to a bird's place in a hierarchy. Personality traits such as boldness, aggressiveness and the ability to explore new situations are all highly variable among individual birds and, remarkably, it has been shown in the laboratory that they are heritable. So, back to your conversation with your neighbours, you can comment upon another successful supplanting attack and say: 'That bird is just like its father.' Personality traits are a buzz subject in animal behaviour at the moment. Some chaffinches, for their part, can apparently be braver than others, returning to a bird feeder more quickly than their flock mates. Others are warier, flying off at the slightest sign of danger when more confident birds stay feeding. Most garden bird enthusiasts who watch their visitors carefully are probably more aware of personality traits than are scientists.

Another sign of dominance mentioned above is body condition. You might expect the dominant birds to be the biggest and best of their generation – after all, females choosing mates almost invariably prefer large-bodied individuals across the board of bird species. Despite this, the dominant birds in flocks are leaner, not bigger. It pays to be lean and not big because bigger birds are slower and a better meal for a sparrowhawk or other predator.

A variety of attributes, therefore, contribute to a bird's place in a hierarchy. But does this matter? It might seem obvious that being subordinate to other individuals in a feeding flock could affect your chances of survival. But does this actually happen?

It seems that it does. In a study of greenfinches, the researchers found that subordinate birds tended day by day to keep higher fat reserves in their tissues than their dominant flock mates did. The assumption is that, if weather conditions suddenly deteriorated, birds with high rank could more easily increase their food intake at the expense of their subordinates, simply because their place in the hierarchy allowed them better access to food during an emergency. If the birds themselves are trying to mitigate against this, then it stands to reason that their rank gives them a lower chance of survival. Having said that, though, hunger will also embolden birds with low social status so that, at times, they can supplant a bird of higher rank. But there is no doubt that subordinates are the most vulnerable members of a flock.

Some long-term studies bear out a relationship between rank and survival, but the effects are small. In the case of the willow tit, a rather scarce bird in Britain that lives in small groups in winter, a bird's rank as a juvenile predicted its overall life expectancy. The black-capped chickadee's survival was more nuanced, with a small increase in life expectancy correlated to its rank after the first year. However, these studies are very difficult to do, because it is hard to measure rank and survival in a wild bird group over several years. What isn't in doubt is that rank has a marked effect on lifetime reproductive success. Dominant birds pair up more quickly, have bigger clutches and fledge more youngsters.

In certain situations, hierarchies certainly can be a matter of life and death. If you don't quite believe this, take note of a study of woodpigeon flocks that reveals just how uncompromising birds can be to their own kind. Woodpigeons, as every farmer knows to their cost, often form large feeding flocks on crops. These flocks typically take the form of an oval. The dominant individuals feed towards the centre of the oval, and the subordinate birds, usually youngsters, forage on the outside – if the flock is moving forward as a unit, they are invariably shunted to the front. The reason that the youngsters are forced onto the periphery is that here they are far more likely to be picked off by predators such as sparrowhawks. When a hawk attacks, it tries to isolate a particular individual and this is easier if the target is already on the edge. Over the course of the autumn and winter, the population of these youngsters buffering the rest of the flock diminishes.

In fact, the subordinates have a double disadvantage. Foraging on the outside, they have to be particularly vigilant as they are constantly exposed to danger. This necessitates endlessly lifting up their heads and checking the horizon, which decreases their pecking rate. Meanwhile, the dominant birds on the inside of the oval have bodies between them and any potential attack from a predator. Their lower risk of predation equates to a less urgent need to be constantly checking around them, and they spend more time feeding and less time being vigilant. Over the course of time, the birds at the centre feed well and maintain their condition, while the subordinates on the outside can lose condition on account of having to spend more time looking up. As they lose weight and health they are ironically more vulnerable to a predator anyway, and birds in this position are frequently picked off. So the rich get richer, while for the poor, things get steadily worse. It's the way of the world and the way of hierarchies in flocks.

You could ask: why don't these young woodpigeons simply leave the flock and join another one? That is what they often do. Unfortunately, though, smaller flocks are more vulnerable to predation anyway, so any pigeon has to weigh up carefully the option of defecting or not.

At this point, you might be tempted to ask what the point is of being in a flock, especially if you are at the bottom of the pile. Such a thought is valid, but in the wild, birds answer the question decisively. Think how many songbirds gather in flocks, in the winter at least. It isn't just birds such as chaffinches, greenfinches and rooks, but also blue tits and great tits, yellowhammers, skylarks, starlings, meadow pipits, sparrows and redwings. A high proportion of birds flock. It simply has to be worthwhile – actually not even worthwhile, but essential.

So what advantages are there in flocking? The most important advantage is undoubtedly communal vigilance. The adage of having more pairs of eyes to spot trouble works well, and when birds have any kind of trouble, it is usually life-threatening peril. A range of predators will be as eager to kill and eat a small bird as the birds themselves will be to avoid death.

To this end, birds have adopted signals to communicate danger, or potential danger. These are known as alarm calls, and they are unusual in that they can usually be read and understood by the community at large so, for example, a great tit can respond to a long-tailed tit's alarm call. The 'hawk-alarm calls' are on a similar pitch and have a drawn-out nature, such as 'see, see' or a trill.

Hawk-alarm calls are so distinctive that humans can recognise them, and I have had hours of entertainment as a field trip leader saying, 'A sparrowhawk should be up in the sky above us right now,' on hearing an over-long trill from a treetop blue tit. Often the predator summarily appears and people are highly impressed. But anybody who spends time in a British woodland can quickly learn to recognise a state of general alarm in birds for themselves. It isn't so different from detecting a fire alarm.

Some birds appear to violate the boundaries of other species in an effort to benefit from their vigilance. On southern English heathlands, a rare bird dwells called a Dartford warbler, named after Dartford Heath in Kent. With a body the size of a wren's, and with the same skulking habit, you might overlook it completely were it not for the Dartford warbler's long tail, almost as long as the rest of the body, which it frequently holds cocked up when it makes a brief visit to the top of a gorse bush. The Dartford warbler spends much of its time in gorse or heather, where it forages for small insects throughout the year.

Delightful though the Dartford warbler is, it is also a stalker. Ask any birdwatcher what the easiest way to see one might be. They will tell you: find a stonechat and there will probably be a Dartford nearby. And it's true. I have done this countless times. Stonechats are easy to find; they are related to robins and stand sentinel on the tops of bushes and trees, looking down onto open ground for terrestrial insects. Stonechats fly from bush to bush, to change their perspective. They are large-headed, short-tailed birds with an orange wash to the plumage. The male is handsome, with a dark head, white collar, white on the wings and almost reddish breast. You cannot miss him, and neither can a Dartford warbler.

Dartford warblers don't necessarily compete with stonechats for food, but on occasion, they do follow them everywhere. The reason would appear to be that stonechats are extremely vigilant. Their habit of watching for food from a high perch makes them experts at spotting danger as well, whether it be an aerial danger, such as a hobby, or a ground-dweller such as a stoat or a snake. The stonechats are forever alarm-calling, and this benefits the Dartford warbler. The Dartford warbler uses the chat as a minder, just as birds moving around in flocks benefit from their flock mates' vigilance.

However, this is not a relationship of mutual benefit. Researchers have found that stonechats change perch more often when Dartford warblers are creeping at their feet. It seems to impair their foraging ability, and they often fly considerable distances in an apparent effort to shake off their stalkers. The arrangement is good for the warblers, bad for the stonechats and excellent for those birdwatchers who are trying to find a rare, skulking prize.

Predator avoidance is not the only advantage to flocking. While more pairs of eyes are good at spotting danger, they are also potentially better able to spot food sources. If a bird is feeding on insects in the summer, which are virtually dripping from the canopy, it doesn't need any help. However, if it typically feeds on a scarcer food source, or one that is available in patches, it is likely to benefit from flocking. A classic example of such a food source is seeds. Seed-eaters are often specially adapted to exploit certain kinds of seeds, and these are often sparingly distributed over a wide area – a good example would be the thistles and teasels favoured by goldfinches, which are found in fields and edges rather than in woods. Certain plants, or even patches, will not all come into seed at the same time; instead, they form a rich but ephemeral opportunity. It can, therefore, be challenging for birds to find these, and in such conditions gathering into flocks often helps.

The goldfinch's diet actually affects its nesting behaviour. Or, strictly speaking, the diet of its chicks affects a goldfinch's nesting behaviour. In contrast to a large majority of songbirds, including chaffinches, goldfinches mainly feed their young on regurgitated seeds, as opposed to insects. While insects are easy to find just about anywhere, seeds, as we've seen, are patchily distributed and often found over a wide area. As a result, goldfinches don't maintain any territory beyond the intimacy of their nest-site. They don't need to sing for long periods of time and they don't endlessly bicker with strangers. Released from this burden, they are free to join other birds foraging. That is why you tend to see goldfinches in flocks right through the summer.

There is one third potential advantage to flocking, which is often overlooked and very hard to measure, and that is the transmission of foraging techniques by birds observing each other. This could happen, especially, in flocks of young birds that are wandering around having left the parental territory. While every bird has an instinctive idea of how to find food, some individuals are simply better than others at foraging. By seeing how their peers find food, young birds could learn new tricks of their own. Imagine, for example, that a blue tit has never tried looking for food among dead leaves. If, during a time of foraging communally, it noticed a flock mate successfully extracting food from a bunch of dead leaves, it might be encouraged to do the same, and thus expand its repertoire of places to look. There must be endless opportunities to learn from other birds, and even from other species. Later in the summer, youngsters join flocks of adults; surely they will pick up some more tips from the experienced birds?

While the flocking habit is a given for many birds, it does, of course, come with the headache of those all-pervading hierarchies. I hope I do not offend anyone by drawing parallels with the caste social system among people in the Indian subcontinent; I am only saying that, for the lowest castes, being at the bottom is an everyday problem and an everyday irritant. In the same way, birds have to endure their position and are constantly reminded of it. Often, they cannot escape it.

Take the magpie, for example, everybody's 'favourite' garden bird (see chapter 5). Looking at a magpie, you wouldn't expect it to experience much angst unless you have witnessed its running battles with carrion crows, the rage of which seems to burn brightly as long as the birds' hearts are beating. But the magpie is burdened by living in a stratified society, with a sharp division between the haves and the have nots. In many a magpie society there are not enough territories to go around, meaning that some magpies are in possession of one and the rest are not. The rest don't like it. At times, the tension boils over.

As you might expect, fights over territory take place early in the breeding season. The skirmish isn't particularly violent, but it is surprisingly public. Many magpies that don't have territories spend time in loose associations with other magpies; indeed, flocks may have several pairs in them. And when a pair challenges the owners of a territory, this happens in a flock context, like a demonstration. It is known as a spring ceremonial gathering.

Many birders have seen this going on. From the outside, it just looks and sounds like a noisy magpie conflagration, with anything up to twenty birds involved. (Incidentally, magpies can be noisy at their roosts, but what we are talking about here takes place in daytime, not close to bedtime.) The birds chatter in their scolding way, and there is much fluttering about and chasing. It is hard to know what is going on.

Essentially what does happen is that a younger pair, upstart kids on the block, simply stage a sit-in at the edge of another pair's territory, usually with other flock members in tow. Not surprisingly, this incites the ire of the resident male, which confronts the interloper with angry calls and physical attacks. While these two indulge in a shouting match and chasing, those birds not directly involved seem simply to perch and watch the fun, calling noisily. It is difficult not to think here of a human playground fight when everybody gathers around to encourage the two protagonists to slug it out. However, this can, in theory, deadly serious – the loss of a territory would be a disaster to a pair, possibly costing them any further breeding attempts in their lives.

In practice, the invading male makes his point and is eventually evicted, with much noise and commotion. Sometimes, though, he can gain a foothold on the edge of a territory and then, little by little, expand it. In bird society, possession is probably even more than nine-tenths of the law.

In birds, hierarchies begin in the nest. You might think that every parent would want its chicks to have an equal chance of survival, and would feed them accordingly. But that is falling into the trap of assuming birds act like humans, and they absolutely don't. That attractive bunch of young nestlings that you can see in your bird box if you have a nest-cam are not a brood of equals, some are more vigorous than others. When a parent tit enters a nest with a caterpillar, it doesn't give it to the youngster that most needs it, but to the youngster that begs most loudly and insistently. When it enters the next time it does the same thing. The upshot of this is that, in terms of body weight, the rich get richer and the poor get poorer. The already vigorous keep well supplied, while those on the margins, perhaps born weaker or slightly later, beg less well, don't put on weight and become less and less adept at begging. The only light in this dark picture is that nestlings quickly become satiated, and fall contentedly asleep. This gives the weaker nestlings a chance of survival.

Some species of birds actually use inequality in the nest as a deliberate strategy. It is technically called 'brood reduction' and it is horrific, at least in our eyes. Brood reduction occurs in species with a highly unpredictable food supply, including such familiar names as swifts and jackdaws. These birds may lay multiple eggs but, in contrast to blue tits, chaffinches, blackbirds and your other average British songbirds, they begin incubation before they have finished their clutch. A clutch of blue tit eggs, even when there are fourteen of them, all hatch at about the same time, on the same day, because the female doesn't begin incubation until she has finished her clutch (more or less, anyway). This means that each chick is about the same size and has a roughly equal chance of survival at the start. In brood reduction, they don't even get that chance.

Take the case of the jackdaw, a bird that will be familiar to many readers. This member of the crow family lays three to five eggs, but rather than laying one a day the female lays one every two or three days. She will then usually begin incubation with the second or third egg. While this might seem to be a small detail, it means that in a brood of four, incubation for the last egg could begin a week after the first and, because each egg takes about the same time to hatch, the last nestling appears seven days after its oldest sibling hatches. This means, of course, that it will be relatively tiny compared to the rest of the brood and could never possibly hold its own when begging for food.

The point of brood reduction is this. If food is abundant, the older young will be easy to feed well, and while they are dozing with full tummies, this gives the runts in the pack a chance to be fed and to survive. The real advantage, though, comes when food is scarce. In these conditions, brood reduction ensures that the older chicks get the first pick of all food to be brought in, and thus a decent chance of survival. There will rarely be enough food to satiate them. The younger siblings won't be able to out-compete them, so they will basically starve to death.

This might seem cruel, and of course, it is. It is raw nature. But remember, the adult jackdaws don't know whether food will be scarce or plentiful when they lay their eggs, which take seventeen or eighteen days to hatch. By staggering the hatching times of their young they are hedging their bets, although being a hedged bet is a tough fate for a youngster. The brood-reduction strategy ensures that, even when food is scarce they have a good chance of raising one strong chick. If all their chicks had equal begging abilities when food was scarce, it would be a case of not enough for anyone, and the whole brood would probably be weakened – or lost.

In the case of the jackdaw, there is a rider to this grisly plotline. Male jackdaws are larger than females, and the male youngsters are more likely to starve in poor conditions because they have higher energy requirements. Studies suggest that most late clutches, laid when the jackdaws' food supplies were beginning to dwindle, are significantly biased towards females, especially if there are only a couple of eggs. It seems that jackdaws are able to produce female chicks almost to order, bearing in mind they have a better overall chance of survival.

Brood reduction is not pretty. It is, perhaps, the least subtle of all the hierarchical arrangements in bird society, with one of the most rapid outcomes. But it is only one form of competition in which birds are immersed, and only one type of way to die at the hands of rivals. If ever the saying 'survival of the fittest' was appropriate, it is in this aspect of birds' lives.

CHAPTER FIVE

Death and Declines

I was idly watching some house sparrows out of the window of a villa in the south of Spain. I had spent the day revelling in the marvellously rich birdlife of the Coto Doñana, and now it was time to relax. The sunshine was beginning to lose its strength, and the sparrows had come out to play. Several came down to the baked soil to preen and dust bathe, while others had found a small pool in the corner of the lawn, by a sprinkler, and were water bathing. The house sparrows' downtime was, as ever, sociable, with much excited bickering and chirping, over nothing in particular. Only house sparrows, teenage humans and radio DJs are capable of spouting so much profitless inanity.

There is a charm about the house sparrow, though, a cheeriness that softens its vices as a species: scruffiness, familiarity and cheek – if the house sparrow was a human being it would sell second-hand items with a straight face. House sparrows might be the loose change of birding, but we appreciate them nonetheless. There is always something happening in a house sparrow flock. If you watch a warbler for half an hour, it might not do anything else except flutter around catching insects, but house sparrows aren't like that. Even if all they are doing is feeding, fights will still break out, or they will fly to and fro, casual and restless.

The sparrows had chosen a small patch of loose soil at the base of a small bush for their dust-bathing activities. Dust bathing is quite an unusual activity among birds, and nobody seems quite sure why they do it: perhaps the dust simply grinds parasites off the plumage or otherwise cleans them? Whatever its purpose, house sparrows do it a lot, and it is difficult to watch them without thinking that pleasure is involved: the birds squat in the dirt, fluff their plumage and roll around, using an identical action to normal bathing. The best dust-bathing spots are disputed, and you can imagine Spanish ex-pat sparrows leaving their towels to reserve a place. On this occasion a queue of sparrows formed, with the waiting birds sitting on perches in the bush, excitedly chirping.

Quite suddenly all the sparrows in the bush flew off in a panic, and there was silence. A moment after the mass scatter some individuals returned to the ground and a communal cheeping began, with greater intensity than before. One or two individuals flew hesitantly towards the bush, angrily calling before flying back again, and facing towards the vegetation. At first, I could not make out what had happened; the panicked reaction was similar to a sparrowhawk strike, but there was no way that a medium-sized predator could have escaped notice. Likewise, no cat could have crept in. It was only when I looked more carefully into the bush with my binoculars that I realised the cause of the commotion. Wrapped around the trunk and lower branches of the evergreen bush was a plus-sized grass snake, and clamped firmly in its jaws was a still-struggling sparrow.

The following moments could have come from a television documentary as the film reveals something you cannot stop yourself watching, however ghoulish. The snake gradually manoeuvred its malleable jaws centimetre by centimetre over the sparrow's body, as its prey's flapping gradually weakened. The other sparrows, all now aware of the snake, mobbed it with continued loud calling and mock attacks. They weren't trying to rescue their colleague but to remove the predator from their territory. It was nothing but a flimsy, futile gesture. This particular snake wasn't going anywhere. Eventually, the sparrow's body disappeared down its throat and the reptile sat tight to digest it in the remains of the evening sun.

As I watched this drama unfold I was more impressed than upset. We tend not to think much about snake predation in the UK, and have very little sympathy or empathy for frogs, which are far more frequent prey items of snakes. On this occasion the sheer privilege, if that is the word, of seeing such an unusual happenstance shielded me somewhat from the reality of the drama. In some ways, too, the sparrows' angry reaction felt like enough of an appropriate response.

I can only remember one occasion when I found a predatory act truly upsetting. The victim, in this case, was a kittiwake, a member of the gull family. Kittiwakes aren't exceptionally pretty or cute, they aren't rare and they're not particular favourites of mine. However, the death of this kittiwake was strangely and hauntingly unpleasant.

The kittiwake, an adult, was standing on a sandbar along with a number of other gulls, of several species; it was a scene I have witnessed countless times before. Nothing much was happening; it was the end of the breeding season and the birds were simply loafing around, as gulls often do. However, the kittiwake was standing next to a great black-backed gull, not yet in adult plumage but still a brute of a bird, at least twice the kittiwake's bulk. Without warning, the great black-back suddenly grabbed the kittiwake by the neck. I assumed at first this was simply a bit of bullying, but the black-back did not immediately let go. The kittiwake flapped in discomfort, and on several occasions almost broke free but, agonisingly, it couldn't quite do so. Whether this emboldened the great black-back I do not know, but after at least five minutes it wandered, ever so casually, towards the end of the sandbar, with the weakened kittiwake still in its jaws. It then proceeded to lean down and hold the kittiwake's head under the water, attempting to drown it. This was like the murder scene in a crime movie. The black-back couldn't hold the smaller bird down all the time, but several times lifted its head up again, before the next dunk. The kittiwake's struggles carried on for an uncomfortably long time – indeed, I even had time to think about intervening before realising that there was too much water between me and the birds – but the poor creature eventually drowned. Once it was dead, the great black-back ate it on the spot. The grey-and-white plumage of the kittiwake turned into a bloodied heap.

I came close to weeping while witnessing this cruel spectacle, and it haunted me for days afterwards. In many respects, it was a privilege to see what was another unusual passage of predation, but there was something different about this death compared to the sparrow's death. One aspect was the nature of the predator. The mugging seemed to be entirely random and casual: the gull didn't really need to assault the kittiwake. It just did. Great black-backed gulls are omnivores and have an opportunistic nature. They often eat puffins and other smaller birds, especially chicks, but not normally adult kittiwakes. The kittiwake was 'unlucky', if you will. One moment it was loafing next to the black-back, the next moment it was on the path to death.

Another haunting aspect was different from the sparrow drama, and that was the reaction of the nearby birds. The sparrows mobbed the snake furiously after its strike; they would have known the stricken bird by sight, a member of their close-knit flock. Their alarm and anger were evident. Here on this sandbar, there were plenty of witnesses, but the kittiwake suffered a lonely death. The rest of the gulls did not react at all. They didn't move unless, inconveniently, the struggling birds bumped into them. As the large bird drowned the kittiwake in broad daylight, the other gulls metaphorically kept on reading their newspapers. And – who knows? – perhaps the kittiwake was paired. If so, on a nearby cliff, its partner would never know what had happened. It was a brutal scene of finality, of potential wasted, with indifferent witnesses.

Until that day I never really understood why people got upset when sparrowhawks broke the tranquillity of their garden and took 'their' birds in an attack next to a feeder. It is, of course, all part of the natural process of eating or being eaten – nobody speaks up for the lovely little caterpillars that are eaten by the very same tits that are taken by hawks in the garden. However, since that day, quite a few years ago, I can empathise better with those who would prefer it if sparrowhawk attacks didn't happen.

The issue of predator attacks on songbirds is one of the hot-potato issues in ornithology. It isn't only about sparrowhawks, of course, it also concerns magpies and other birds labelled as 'killers'. On our journey into the dark heart of the reality of life for a bird, the theme of this book, can we reach into the light a little more? Many have tried to answer the various questions. Do sparrowhawks (or magpies) keep birds away from gardens? Do sparrowhawks reduce bird populations? Or, to put it another way, are widely perceived songbird declines linked to widely perceived (and scientifically proven) predator increases?

And should we attempt to control predators' numbers, and if so, why?

Before moving on to examine this issue, there is a tug at my own heartstrings. It is a curious thing, but I feel that the birds themselves are collectively lobbying me. If this book is an exposé of what life is really like for birds, especially songbirds, with as little prejudice as possible, then there is something that the collective avian community would urge me to examine. What is the real truth about birds and cats? I don't really want to touch this issue because I am afraid of being unpopular. I am not sure that I want to know the truth because the current research is not comfortable for cat owners, especially those who also love their birds. But we will look at this towards the end of the chapter. In some ways, I hope you don't get that far.

So what can we say about sparrowhawks? They are specialised bird-hunters, catching their prey using an imperious combination of surprise and agility. Bearing in mind that a living bird will always do everything in its power to avoid being caught and killed, the predators have to surmount great resistance and difficult obstacles. They rarely if ever take the easy option – meat that's already dead (carrion). Sparrowhawks do eat a few small mammals, which constitute on average only 3 per cent of the diet. These include voles, shrews, young rabbits and occasionally larger items such as weasels or stoats.

Sparrowhawks are small predators, and so their diet disproportionally includes songbirds. The weight range of captured birds is 10–400g (0.35–14.1oz), although female sparrowhawks, which are much larger and heavier than males, will occasionally kill woodpigeons as heavy as 600g (21.16oz). However, most prey weighs less than 50g (1.76oz). In Britain and Europe, the sparrowhawk has been recorded taking 120 different species. Extensive research has indicated that your average sparrowhawk needs 53g (1.87oz) of food per day, but in order to obtain this much it has to kill birds up to a value of 82g (2.89oz). All in all, it needs to kill 2 or 3 sparrow-sized birds each day in order to survive, but only 1 pigeon or dove.

If you take a year, and try to work out how many birds a pair of sparrowhawks raising an average number of young would catch and eat, you get 2,000–2,200 sparrow-sized birds each year. Of these, about 1,500 will suffice for the adults. These figures are much reduced if the sparrowhawks were to kill a high proportion of larger birds. Over the whole country, it is thought that sparrowhawks kill 50 million birds a year.

It is a high total, but of course, if we take the figure of 1,500, a single sparrowhawk will not take all of these from one place, particularly not a small garden. Individuals don't have a territory that they defend, but will range 3–4km (1.86–2.85miles) from a nest in the summer, and at least that much in winter. It is impossible to say what an average number of small birds in a sparrowhawk home range will be because there will be enormous differences in habitat and land use.

Do sparrowhawks deplete bird numbers? Of course, by definition they do. On the other hand, it is an ecological principle that everywhere they go, sparrowhawks need enough food to survive. If there wasn't enough food somewhere, they would starve if they remained for long, or hunger would force them to relocate. To put it another way, it is in the sparrowhawk's interest to have healthy populations of its prey species available. In the natural world, there is a balance between predators and prey. The numbers of predators are limited by their prey abundance, not the other way round. If, for any reason, an area suffered a depletion in the number of birds that weigh around 50g (1.76oz), the sparrowhawks would feel the effects most and would disappear first.

On the other hand, sparrowhawks can theoretically deplete the population of a particular species of bird, or in an area. As it is, they have favourites. Birds that often feed on the ground in the open are attacked disproportionately – unsurprisingly, house sparrows fall into this category. A study by Dr Christopher Bell and colleagues from Cambridge University suggested that the recent increase in sparrowhawks may have contributed to the decline of the house sparrow in Britain. The sparrowhawk reached a low point in the 1960s, but its numbers have increased by 29 per cent since then, apart from a small decline very recently. Meanwhile, the house sparrow population dropped by 69 per cent between 1977 and 2010. At the moment the consensus is that the sparrow decline is mainly caused by other things: low survival of first-year birds in farming areas, owing to a reduction in winter food supply, and in urban areas by a reduction in nest-sites and poor food supplies for the young. However, the survey points out that when sparrowhawks recolonise an area, they might reduce a sparrow population that had temporarily become adapted to a hawk-free zone.

Those, then, are the bald facts and basic ecology. It is now time to discuss some widely held opinions. Take this statement, quoted on the public forum birdtablenews.com: 'I keep decreasing the amount of food I put out for the birds but, thanks to sparrowhawk attacks, the little birds are visiting my garden less and less often for fear of being lanced by hawk talons and eaten alive.' The clear implication of this is that the presence of sparrowhawks actually deters birds from entering gardens and using the feeders.

Is this true? We are not talking here about larger birds, such as pigeons or crows, competing with smaller birds and deterring them from visiting a feeding station by 'hogging' access to it, which certainly can happen. The clear suggestion is that small birds avoid coming to bird tables, or even whole gardens, because of the *threat* of predation. They are supposedly keeping away from an area because they realise it can be dangerous.

As far as I am aware, there is no study that can prove or disprove this assertion, because the practicalities of carrying it out are so punitive. You would have to find a garden or neighbourhood with no sparrowhawks in it and then monitor every individual small bird and its movements in your study site. You would then have to introduce a sparrowhawk into this novel area and exactly monitor the movements of every bird previously present, to see how they are reacting. Even if you could prove that the bird population subsequently declined, you would still have to link the decline to living birds actively avoiding the area, as opposed to direct predation. It would be quite a study.

There are, though, many good reasons to take issue with those who fear that sparrowhawks actually keep birds away from gardens. The first reason is that birds don't visit feeders in order to entertain us; they visit feeders to feed. As you might have gleaned from the earlier pages of this book, birds are creatures of compulsion; they must eat to survive, and if it is easy to acquire food anywhere, they will go there.

Another reason is that predation is a constant threat to birds. It takes many forms, not just attacks by sparrowhawks. There is never a time when a bird is not vigilant because there is never a moment when birds are not at risk. When feeding on the ground, for example, birds look up and around every few moments. One of the main reasons birds gather in flocks is that it increases communal vigilance, to make a sparrowhawk attack less likely to be successful. And, of course, sparrowhawks are wide-ranging; they won't spend all day at a single feeder. A bird might avoid a feeding station that is used by sparrowhawks, but it doesn't have anywhere to hide. It could be caught in a field next to the garden, or in a nearby wood. If a bird wished to avoid any threat of predation, it would metaphorically never get out of bed.

If you allow yourself to examine the issue of sparrowhawks by surfing the internet, you will find that many people seem to advocate control of the predators. Every time I see this mentioned, I find it hard to understand. Are these people really suggesting that, in order to mitigate against witnessing the upsetting sight of a sparrowhawk attacking small birds in the garden, you should do to the sparrowhawks what they are doing to the tits or robins – kill them? Let's not beat about the bush; if you want fewer sparrowhawks, you have to destroy them. This solution replaces the entirely natural drama of a puff of feathers grabbed by talons by an entirely more sinister puff of feathers as a wild bird is taken down by a bullet. Is this really what a householder wants? If they care about the death of a small bird, do they also not care about the death of a larger one? Is the human love of birds really so conditional and shallow?

What most people want is merely to avoid the upsetting scene; they don't want a violent solution. There are exceptions, of course. I can perfectly understand why racing-pigeon enthusiasts want to destroy predatory birds; it's so that their much-loved and often expensive birds fly in the fresh air with less danger of being caught. I have sympathy with them when they are honest. However, make no mistake, a racing-pigeon enthusiast wanting to rid the place of sparrowhawks or peregrines could not give a damn about songbird populations. Such people might use declining songbird populations to justify predator control, but it is a con.

One of my own biggest bugbears is when somebody declares that there are 'too many' of something. 'There are too many magpies,' they cry. 'There are too many predators.' 'Too many pigeons.' On what grounds, in a wild ecosystem, are there too many of anything? Away from comfortable households, we know that there are natural checks and balances. It is called natural selection, or 'survival of the fittest'. If there was an unsustainable population of magpies in a suburb, it wouldn't be sustained because it is unsustainable. If there wasn't enough food for magpies and sparrowhawks they would move on or die. If there really were 'too many' pigeons or parakeets, natural selection would wipe out the excess.

'Too many magpies' is simply an example of somebody's opinion. It needn't be based on anything trivial, like facts, for it to be expressed. In political life at the moment, people might say there are 'too many migrants', but on what is their opinion based? Who decides how many is too many? Yes, people are afraid of being overrun culturally, and of stretching limited resources by allowing in immigrant people. But when somebody says there are too many of anything, we must always challenge them.

The magpie issue in gardens, and in the countryside at large, is a far less clear-cut issue than the case of sparrowhawks, although many people would deny this and squeal for magpies to be eradicated from the suburbs. However, the sparrowhawk is a supreme year-round predator of birds of all ages, while the magpie is very much a part-time raider of nests and a part-time consumer of eggs and nestlings. It doesn't attack adult birds, except for very occasional instances which are a negligible quirk. In other words, once a bird has left the nest and is living independently of adults, it is very unlikely to be killed by a magpie. It is much more likely to be run over by a car, killed by a cat or even knocked unconscious by flying into a window. It is important to say at the outset that a magpie eats young birds about as often as you and I might go to a restaurant and eat oysters – which is, I presume, hardly ever, and only in season.

Equally, you cannot exonerate magpies. They are frequently caught in the act of raiding nests, and nobody denies that this happens. Jays, incidentally, probably raid as many nests as magpies do, but they rarely get rumbled because they usually do it under cover of oak trees and the vivid spring canopy. However, observing magpies attacking defenceless baby birds is a deeply unpleasant experience. It is like a slasher horror movie, complete with a soundtrack of distress calls. It often plays out in the open, close to human habitation. The attack can be long and drawn out. When it happens, people observe it disproportionally often.

And when we do see a magpie nest raiding, our human nature always seems to side with the smaller bird, the blackbird for example. Our sense of injustice kicks in as we see the larger bird prevail.

It is surprising, taking a dispassionate view, just how few birds, even in the garden, are under threat from magpies. The most typical nests raided are those of cup-nesting species, such as blackbirds and song thrushes. These are, of course, much beloved by human gardeners, but they represent a small subset of affected birds.

The song thrush is sometimes assailed by magpies, and this same species has famously been in long-term decline in Britain, although it has stabilised recently. The years of song thrush decline coincided with years of magpie increase, fuelled by the black-and-white bird's invasion of suburbia (this possibly aided by the planting of suitable trees for magpies to nest in, and by an increase in roadkill). How could the two things – the 'victim' song thrush's decline and the 'perpetrator' magpie's increase – not be related? This is the perfect scenario for an old-fashioned detective novel. Is it not black-and-white, an open-and-shut case? It's the one where the conventional police have seen the crime, found witnesses and have duly convicted the killer. Then, however, the famous sleuth comes into the picture and, little by little, unpicks the crown's case. All is not what it seems.

I have a recollection of a member of the public acting the same way as the simple-minded cops in that detective novel. At the time I was leading a group of people around Bedfont Lakes Country Park in west London, near Heathrow. We were looking at reed buntings on an early spring day, minding our own business. Along came a power-walker, whose adrenalin had presumably fired up a lack of self-consciousness. Anyhow, he seemed happy to address us, uninvited, as he went past wearing alarmingly baggy tracksuit bottoms.

'There used to be lots of thrushes here,' he said. 'But I haven't seen one for years. It's because of the magpies and sparrowhawks.'

And with that, he sped away. We had no time to say anything. His declarations hung in the damp air.

The alarming thing about this remark was that many people would have agreed with him. The amusing thing about this remark was that, all the time he was speaking, a song thrush was singing loudly from a nearby tree. He was, in every way, the blinkered cop.

I'm quite happy for people to air their opinions. I love it when they do so. If you are upset when a magpie raids a song thrush nest in your garden, I am fundamentally delighted that somebody else cares about birds. However, please forgive me when I remind you that just because it has happened in your garden doesn't mean that it is affecting the song thrush population at large.

What, then, is the problem for song thrushes if it isn't magpies? It turns out that the major recent losses in song thrush populations are caused by problems in first-year survival. The age-class that is really suffering are those that have already run the gauntlet of magpies, survived past the spring and summer, and then are struggling to cope with their first winter. It seems that first-year birds aren't finding enough food.

It may be that magpies don't help. Yes, they take song thrush eggs and young. But they aren't the major cause of the decline. It is amazing how often people just don't accept this. Perhaps they simply don't trust scientists.

There are, however, some studies that reveal a magpie problem. In one study, only 5 per cent of urban blackbirds raised young successfully, and a substantial number of nest failures were down to magpie predation. In the urban and suburban habitats, it is thought that magpies more easily find nests than they do elsewhere, which can undoubtedly cause local effects. Since magpie numbers have increased by more than 100 per cent in the last thirty years, should we be worried?

Not, apparently, if it is overall numbers of songbirds that concern people. Overall, the magpie is constantly and consistently exonerated from blame.

A large study conducted in the 1990s by the British Trust for Ornithology (BTO) presented a very broad view. They analysed the overall nesting success of 15 species of British songbirds during a time when magpies were increasing in Britain by 4 or 5 per cent a year (1966–1986), and they found no appreciable decline in any songbird analysed. In fact, in some areas of woodland, songbird numbers increased at the same time as magpie numbers. If magpies are an existential threat to songbird survival, this simply wouldn't happen.

At the same time, songbird numbers in many areas have fallen substantially while magpie numbers have remained the same. This might be a negative result, but it is a compelling one.

Another study published in 1998 and authored by D. L. Thomson, R. E. Green, R. D. Gregory and S. R. Baillie, and using a very large dataset from thirty years of population counts, came to a similar conclusion. In its summary, it states: 'Our results indicate that magpies and sparrowhawks are unlikely to have caused the songbird declines because patterns of year-to-year population change did not differ between sites with and without these predators.' This is good science, published in *Proceedings of the Royal Society*.

Yet another study, in 2010, examining data collected over forty years by the BTO, once again concluded the same thing. It was sponsored by an organisation with the attractive name of Songbird Survival, which was founded on the basis that predators are to blame for songbird reductions, and this is not widely accepted only because it hasn't been researched properly. To their credit, this organisation is paying for research of its own, but it doesn't help that one of their sponsored papers merely casts doubt on the methodology of previous research. This feels like clutching at straws. The suspicion remains that many of their supporters are keen on shooting, and would rather like a 'scientific' basis for keeping on with their hobby. But some, no doubt, do believe what they are saying.

In 2011, the Game and Wildlife Conservation Trust began a wide scale cull of corvids to see whether any effects could be magically found. It might not be worth holding one's breath for a different result to all the others that have been published saying the same thing. It is hard not to wonder whether certain organisations will simply commission research until somebody agrees with them.

The magpie situation can actually be summarised quite succinctly. Magpies are part-time raiders of birds that build cup-nests in environments where these nests are easy to find. Magpies don't frighten birds away from gardens. Their attacks are upsetting, but despite the recent 100-per cent increase in magpies in the UK, and the simultaneous drop in numbers of many songbirds, the one hasn't caused the other. A cull of magpies might have local benefits, mainly in urban areas, but is completely valueless nationally. The problems for songbirds lie elsewhere.

The important point in this is that there are vested interests, shooting interests, that have jumped on the songbird bandwagon as justification for their hobby. Actually, I am not completely against shooting; I don't like it and wouldn't do it, but I accept that sometimes animal numbers, such as those of deer, need controlling. What I really don't like are liars. Hunting and shooting have a long heritage in this country, and maybe some feel their traditional sports are in retreat. But by breaking the law at times – as they have done when controlling hen harriers, for example – and by trying to make us believe that controlling predators helps songbirds in their droves, some advocates of these sports undermine their credibility.

Having said all that, I want you to know that if a major study does, one day in the future, confirm that magpies are, or have become a real menace to wild songbird metapopulations, then I will change my position. I am not afraid of the truth and neither, I trust, are the conservation organisations such as the RSPB and the BTO. If you have read this book so far you will have noticed that there are many scientific studies quoted. They are trustworthy because their authors have no axe to grind and they conducted their research without prejudice. If a study of comparable weight tells us that we should control magpies widely, then great – bring it on.

Here's a question – if you love songbirds in your garden, would you own a cat?

Here's another one – if you advocate culling magpies to protect the songbirds in the garden, would you back up your love of birds by not having a moggie? Or by allowing a cull of feral (not domestic) cats at the same time as a magpie cull?

I cannot count the number of intelligent people who have said to me personally that 'their' cat doesn't catch birds. Some of them are probably right, but I have a feeling that they can't all be. In the cat-owners' favour, research by Rebecca Thomas and her co-workers at the University of Reading did show there is marked variation in the numbers of prey returned to their owners by cats, and only 22 per cent of cats brought back more than four items in a year. The same survey found that a high proportion of cat owners don't think that their cats are a problem, and that many are very resistant to cat management to ameliorate potential predation. More than two-thirds of cats in the survey were allowed to roam freely by day and by night, and fewer than a quarter were fitted with a collar and bell. Although some owners were afraid to fit a collar because they feared for their cat's safety, the main reason for not managing cats is because people aren't convinced there is a problem. They don't realise that, just because their cat might only occasionally bring in a dead animal, they are witnessing the tip of an ecological iceberg.

Although cat predation is patchy and difficult to witness, recent research does suggest that cats could be a serious problem to songbirds in this country. In the same survey quoted above, the estimated numbers of blackbirds, blue tits, great tits, dunnocks, house sparrows and robins killed by cats in selected areas was greater than the proportion of adults on 14 of the 36 randomly selected 1-km^2 (0.38 square miles) sections of the town of Reading. The very obvious conclusion was that the cats could have been causing the birds' populations to decline.

Another survey for the Mammal Society, based on questionnaire samples, estimated that there are 9 million domestic cats in Britain. Based on the returns, between April and August 1997, British cats killed 92 million prey items, of which 27 million were birds. That would be 55 million in the course of a year. The cat is by far the most numerous mammal predator in Britain, and about one-fifth are feral, so are not controlled by people at all.

There is one aspect of cat predation that makes the mammals a very different proposition to sparrowhawks and magpies. Magpies concentrate nearly all their effort into an omnivorous diet and their predation of birds is mainly aimed at eggs or nestlings. Sparrowhawks are full-time bird eaters, but they are also specialists, who depend for their existence on a good supply of wild birds, and their numbers are regulated by their prey species, not the other way round. Cats belong in neither category. Most are fed by householders, so their populations are not regulated by the populations of their favoured prey species.

Another interesting line of recent research has highlighted the non-lethal effects of cats. When a cat or cats are near to a nest, for example, parent birds pay fewer feeding visits to their young than they otherwise would, potentially affecting the overall bird populations. In common with cat predation rates in general, this will benefit from further investigation.

The RSPB maintains that, in common with magpies and sparrowhawks, songbirds probably are able to sustain this level of predation by cats and they don't advocate any cat control. This is probably sensible, and certainly fair. However, more research is needed to see whether the problem is greater than anticipated, because the picture is not as clear as it is for sparrowhawks and magpies.

However, if we are going to be fair on cats, we should be fair on magpies and sparrowhawks, too. If you are a cat owner who would like to see control of magpies and sparrowhawks, you are on the shakiest of shaky ground.

At the very least, cat owners who like birds ought to take the problem of their pet's predation seriously. In the USA, most people apparently keep their cats indoors, which would have a significant impact here. It seems just lazy not to buy a collar, too.

The problem of predation on songbirds is riddled with misinformation, prejudice, lack of education and stirred feelings. In recent years, it is encouraging that there has been much research done, and the penny is perhaps beginning to drop on what the real picture is. Maybe the future is a little brighter.

The birds themselves are immune to the argument. Being caught and eaten is just a part of their lot.

CHAPTER SIX

Repose

As a birdwatcher, I am accustomed to finding gold in unlikely places. I once heard a pied flycatcher calling in my garden when I was putting out the bins, a job I loathe. No rubbish, no pied fly. Now here I am putting petrol in the car at the Sainsbury's garage in Ferndown, another tiresome chore, and something odd is happening.

It is a cold winter's day, and a late afternoon chill blows across the forecourt. Garage forecourts are cheerless cold concrete deserts most of the time, and you half expect tumbleweed to ride the gusts past you as you keep your eyes on the pump meter, the only exciting thing to watch. These days some garages have taken to pumping music over their loudspeakers, making the misery much worse. Wind, cold, concrete and Miley Cyrus is too much for anyone to bear.

This time the forecourt was music-free, and I soon heard the familiar 'chiss-ick' of a pied wagtail calling (we used to call it the Chissick Flyover when I lived in west London – ho ho!). Pied wagtails are among the few birds in Britain that cope with busy, tarmac-covered areas such as car parks and garages, picking up tiny insects from the ground, apparently immune to human business. This bird was running about at the foot of the opposite set of pumps, busily picking at the ground and eponymously wagging its long tail. Pied wagtails are the sort of birds that are almost smart, but never quite immaculately turned out. They have too many messy patches on their plumage.

I heard another call and saw a second pied wagtail flying upwards towards the roof of the paying booth and shop. I didn't think much about this until glancing to where it must have been going, and there, lining the rooftop, silhouetted against the darkening sky, were a whole cluster of pied wagtails, like swallows on a wire. I didn't count them, but there must have been at least forty. There was a small patch of grass on the edge of the concrete, and here more pied wagtails had gathered, at least another five or six. On the roof of a car wash there were suddenly another ten. The scene resembled one of those creepy movies where the hero enters a room, turns on the light and there are suddenly twenty dolls, wearing their blank but sinister expressions, where there hadn't been any before, and it's the same in the kitchen and the lounge. Here I was surrounded by pied wagtails, appearing out of nowhere. Beyond the gloom you could be sure there were more out there, waiting for me.

What I had witnessed is known as a pre-roost assembly. Pied wagtails, in common with a number of other songbirds, gather together to spend winter nights in a community. In contrast to some of the other birds that do it, these characters seem to be drawn to flat roofs, similar to those that you find on top of supermarkets and railway stations. And here they spend the night as if bedding down in one large dormitory.

Some of the most spectacular things that any birds do are associated with their sleeping habits – think of assemblies of starlings, corvids, or even gulls. This is a little odd to us because, on the whole, we don't tend to do much exciting stuff around sleeping, unless you count sex. But birds most certainly do, and their routines are often quite easy to witness. In this journey into the dark heart of bird behaviour, we are going to enter into the heart of dark behaviour.

To be honest, scientists know little about bird sleep. We know that they do sleep, although the details differ between species. I have only seen birds actually sleeping a few times, usually when I have been spotlighting at night for mammals. A bird will sometimes be spooked away immediately, but occasionally it will remain as still as an exhibit in a waxwork museum, clearly fast asleep and doubtless very vulnerable. Many birds spend time in a state which is more resting than sleeping, and will open their eyes at regular intervals.

This chapter, though, is mainly about bedtime routines and, in the case of the starling, about waking-up routines as well. The latter is arguably more remarkable than the former.

Before describing some of the more extraordinary roosting behaviour, what about those birds that actually don't make any hullabaloo at all? There are many examples of these, not least great tits and blue tits. Away from the nest, both sexes use their own roosting holes or crevices, which they treat as their own, although most individual birds have several hideaways in case they are disturbed. These birds just need somewhere that is safe from predators and with some protection from the elements. They enter the roost unobtrusively in the evening and emerge at first light, or just before. They wake up separately, and although later in the day they may well join up with a flock of other tits to forage and feed, at the end of the day everybody goes their separate ways.

Some of our most familiar birds sleep closer together than you might think. Take blackbirds, for instance. One of the most familiar sounds in the garden on a winter afternoon is the loud 'chink' call made by blackbirds as they bed down. It sounds testy and aggressive, and so it is. Blackbirds are highly strung creatures, and the necessity to sleep in fairly close proximity to their neighbours brings about a degree of stress.

Some birds are forced to huddle together in bodily contact in extreme external conditions, usually of deep frost or snow. Wrens have become well known for this, although other birds do it as well, including goldcrests and even robins. Wrens and robins are hot-tempered at the best of times, and roosting communally must be very stressful for them. Of course, if there is a choice between an uncomfortable night and death, they are obviously going to opt for the former.

Wrens have a small body size, which unfortunately means that they lose heat very easily owing to their high surface-to-volume ratio. Of course, they always lose heat easily, but in the daytime they can replenish any heat lost by eating food and burning it off. At night they normally mitigate against heat loss as best they can by finding a crevice in which to sleep, and nine times out of ten they can cope. However, wrens presumably know instinctively when it is simply too cold to roost alone. The laws of physics also allow for such birds to increase their surface-to-volume ratio if they cuddle up to their neighbours in physical contact, so these birds actively seek out company in the evening.

What seems to happen is that a local wren with a territory makes loud songs in order to summon neighbours to the wren sleepover, and it also makes flights around the area.

Word evidently spreads well, and individual wrens are known to come from as much as 2km (1.24 miles) away to take part in the roost, presumably knowing of its existence in advance. Roost sites are always cosy and well insulated in such places as nest-boxes, roof thatch, the nests of other birds, and so on. There are rarely more than 10 birds in a roost, and they settle in rows, squatting several deep with their heads facing inwards. Adult males, adult females and first-years of both sexes may all join in.

Fascinatingly, however, it seems that not everybody is invited. The territory-holding wren has been observed acting as an effective doorman, and apparently refusing entry to some of its neighbours. Why it should do so is not known; is it because their face doesn't fit for some reason, perhaps as a disliked territorial neighbour? One would imagine that the consequences of being evicted from the roost, or not let in, could be severe unless there is another roost nearby. Arriving at a potential roosting site to discover you are not welcome could also be costly, in a situation where any energy needlessly expended could have serious ramifications. However, sometimes the tables are turned, and the territory holder botches the eviction, finding itself forced out into the cold instead.

Another bird famous for huddling is the long-tailed tit. In contrast to wrens, long-tailed tits routinely huddle each night, although if it is really mild they will simply stay close. The stress buster would seem to be that long-tailed tits roost as a family group, not as a collection of disparate and desperate individuals. In a long-tailed tit huddle, which perches together along a horizontal branch, usually low in a thorn bush, there will be a 'senior' pair together with their offspring from last year's breeding season, which could number anything up to eight birds. In addition, there are likely to be adult relatives of the senior male, usually brothers. The situation that first-winter long-tailed tits find themselves in is unique among small British songbirds. In almost every other instance, including other tit species, robins, blackbirds, chaffinches and starlings, juveniles remove themselves from the territory of their parents as soon as they are independent and they may wander some distance away, never to return. However, in the long-tailed tit's case the juveniles, particularly the juvenile sons, remain by their parents' side for many months.

But the very reason that the offspring do not leave their parents' side is because of the urgent need to cuddle in the winter. Long-tailed tits need huddling partners, or they die. Adult long-tailed tits often help at the nest of their siblings in the summer, and it has even been suggested that this behaviour gets them into the good books of their relatives and allows them entry into the huddle for the autumn and winter. Whether this is the whole reason or not, it does demonstrate just how important huddling is.

The actual social dynamics of long-tailed tit huddling groups illustrate the central theme of this book, the completely unsentimental dark side of bird behaviour. Remember, the flock huddles along a horizontal branch. And that means, of course, that some birds will be cosily ensconced in the middle, while the booby prize is to take an outside berth with half your body exposed to the elements. You might think that, since much of the flock is composed of their progeny, the senior pair might allow their offspring to take the middle places. That is what a human parent would do to protect their children. But it isn't what long-tailed tits do. As with any bird flock, there is a hierarchy, and the senior birds take the best positions. The fact that their offspring might die as a result of their extra exposure to the cold does not seem to faze them.

The dynamics of wren and long-tailed tit huddles are cruel but they are certainly logical, with a readily understood explanation. The reasons behind the mass roosting gatherings of starlings and crows are far less easy to understand.

Starlings are rightly famous for their habit of gathering together in the late afternoon to roost, and to visit a large congregation is an authentic natural wonder. Having said that, a modest-sized roost is pretty impressive, too, so you shouldn't be put off if there are only a few thousand birds in your local area. If you arrive at least half an hour before sunset, and choose a reasonably light or sunny day without too much wind or rain, you will find out what the fuss – as proclaimed by a succession of purple-prosed wildlife documentaries – is about.

In common with that other easily accessible avian wonder, the dawn chorus, the arrival of starlings starts slowly, builds up and eventually knocks you off your feet. Every stage is fun, from the anticipation at the beginning (often shared with other people who have come to watch) to walking home in the winter dark afterwards, happily delighted with what you have seen.

The roost I have experienced most often is the largest in Britain, in the Somerset Levels. It is at its best in deep winter, in December and January; after this point, not only are there fewer birds, but they have a habit of changing their roost site without warning, which means that, on certain days, all the action is elsewhere and you won't see a single starling, even though more than a million of them might be just a few kilometres away. There have apparently been up to 8 million starlings here in the recent past, although having seen the roost, I simply cannot see how you can count 8 million birds. Nevertheless, there are plenty.

Ham Wall RSPB reserve, where the roost is most frequently situated, provides a grand stage for a birding spectacular. Although only recently restored to its best from old peat diggings, it has the appearance of an ancient landscape – there is something elemental about the vast ranks of tall reeds, dykes, cold-looking lakes and stark trees. And you see Glastonbury Tor in the background, for a frisson of ancient paganism. The area is very open and, if it was up to me, there is no way I would sleep here, with the potential for deep, hard frosts and icy winds. However, for a tough starling, especially a Russian one, the reeds provide relatively safe sanctuary and their firm stems provide an awful lot of perches – presumably a requisite for a large gathering.

In this place, there are always high-quality sideshows to the starlings – you routinely see great egrets, bitterns and many ducks, and on a number of occasions an otter has raised its blunt head above the water, quizzically observing the pre-roost assembly of humans. The half-mile walk allows you to get the feel of the place, the day and the weather, usually too late to change the fact that you have forgotten your gloves and remembered that the temperature will soon plummet when the sun disappears.

The birds always seem to take longer to arrive than you expect and, as the light fades, you sometimes mistakenly get excited over flocks of finches, or even pigeons. The odd group of starlings might fly over, too, causing your heart to miss a beat. But the sight of the first flock comes quickly. They are always quite far off, on the horizon, and they always disappear behind the trees, teasing you. There are never that many, and you begin to doubt them. Then a more substantial group appears, of several hundred birds. At the moment the flock is flying low and determinedly seems to be heading somewhere else, out of sight. Any minute, now, though, a really large group appears, often taking you by surprise, coming from behind, completely silently, for example. The starlings are no longer starlings – the bird familiar from its scruffy appearance and swaggering walk across your lawn has changed into a dot in a living multitude, no longer itself, part of something bigger, something out of control and perfectly formed at the same time. The flock doesn't fly, it swirls because aerodynamic rules seem to have been suspended.

Sometimes, the big swirl moves away and over. But soon comes the moment when every horizon is busy. The members of big starling gatherings come in from every direction, and some flocks have come from 30, even 50km (18–25 miles) away, just to take part in this festival of sleep. The starlings are still flying low – indeed, they always stay lower than you would expect from watching the wildlife documentaries. On the horizons they are streaming in, often in bursts – a dense flock, then a line, then a dense flock – a living heart tracing in the sky. All the time there is a vocal silence – the evening non-chorus – leaving only the sound of wings, a gentle thundering. Many of the subgroups of the roost make a pass over the reeds, zooming, moving faster than they need, only to wheel away and over back towards the horizon, going nowhere except somewhere else. We know, though, that they will be drawn back any minute – with all the activity, this is the chosen roost site for tonight.

Always there is that moment – a truly thrilling moment – when you thought you'd seen the biggest concentration, were happily impressed and then you are caught by surprise and turn around and something still bigger rolls over. You find yourself laughing at the enormity of a starling wave. By now the birds will be making their own light, darkening then lightening the gloaming, creating a living canopy overhead. Now there are perhaps a million extras in your performance and, if you are fortunate, they will now play their party piece. This doesn't happen every night. Sometimes, usually in damp, cloudy conditions, they don't make any more of a song and dance than low-level swirls. When it is sunny and still, however, they can dance in the reddening sky. This takes birding into a different level of entertainment. It turns hardened twitchers – birders that would kill for a new species on their list – into artists. This is the National Gallery of wild birding.

There is nothing like a large mass of starlings, a murmuration, to write wonders in the sky. Many authors have described them as looking like a moving amoeba, or a swarm of bees. What is always obvious is that there are areas of light density and dark areas of high density which, as we will see, are important. These constantly shift and produce shapes of all kinds – dolphins, people's faces – which appeal to our imagination. But it isn't the picture painting that really makes a starling roost such a compelling phenomenon. From a purely human point of view, it is the sheer collective joy of it, the creating of one thing by millions of autonomous bodies, the sense that there is no end to what the birds can do. The swirling murmurations are spectacular, but it is their spontaneous unpredictability that excites the soul.

The wheeling and convoluting don't last long. Soon the bodies rain down (and they really do) from the sky, and they begin to chatter excitedly as birds evidently decide where they will rest the night. A few years ago I hid in some bushes as the birds settled in, and I felt completely surrounded, and almost a part of things. The strangest thing is that, having largely kept their mouths shut during the aerial display, they now natter incessantly, probably about nothing in particular. The overall impression is that they become starlings again.

The remarkable nature of the starling flights around the roost has, not surprisingly, attracted the attention of scientists asking the question: how do they move around like this in the air and not bump into one another? You can see the possible human applications of any answer – air traffic control, driverless cars, and so on. The funny thing is that in the morning, when the starlings are leaving the roost, they do often bump into one another. In the evening, however, it seems effortless and incident-free.

So how do they do it? The research is still in its fruitful phase and there may be much more to come, but one conclusion is that each individual filters out the overall experience of being in a large flock and only reacts to the movements of the seven nearest birds to it. This helps to explain how the birds manage not to have crashes. Reading that, it all seems possible and you suddenly think 'I could do what a starling does after all'. One assumes that all the birds cooperate and that the mass doesn't allow for renegade white-van starlings to ignore all about them.

The newest research takes issue with this simple solution and suggests instead that starlings do observe the whole flock around them, in terms of how much light reaches them through gaps in the flock. The idea is that each starling aims for something called marginal opacity, which is the optimum density of birds at which they can gather data on their surroundings. Although every bird is aiming for this ideal, some birds are in dense clusters and others are more loosely clumped. The pattern of dark and light, created by dense parts and loose parts, helps the birds to see where they should be going, and they alter their position and angle in order to reach optimum density. Thus the flock appears to be regulating its density by moving around in this dynamic fashion.

When you do witness the aerial convolutions of a large starling roost, there is always one other question that nibbles at your mind. Why on earth do they perform this stunt? And this provokes a broader question: why do starlings gather in large groups to sleep anyway?

The most compelling theory for the existence of the spectacular pre-roost aerial manoeuvres is that they are a communal form of advertisement. Starlings obviously benefit from being in large groups or they wouldn't roost communally at all. And it seems that big roosts are desirable because, again, if they weren't the birds would only ever gather in small groups. In the countryside at large, there are usually ample places – scrubby areas, buildings, reeds – where starlings could roost if they wished. But they choose to gather in a few places in large numbers. If you follow this argument it stands to reason that the roost members want to attract more birds in, because it benefits them.

Other birds also perform visual displays at roosts, especially rooks and jackdaws. The latter rival starlings in the exuberance of their aerial manoeuvres in the evening sky, although their flocks are not as dense and the flying birds look more like lumps of ash from a bonfire caught in the wind. Jackdaws soar and wheel, but they also permit themselves dives at breakneck speed and more flying solo. Pre-roost jackdaws and rooks are also much noisier than starlings, the high-pitched hiccoughing 'jack' calls of the jackdaws mixing with the throaty cawing of their larger counterparts.

The fact that jackdaw pre-roosts are noisy has parallels with some other advertising. It is thought that pied wagtails have special roost 'summoning calls', and wrens may attract visitors by being noisy (not that you'd notice, as wrens are always noisy). However, the same rule applies: birds advertise for roost mates because it pays them to do so.

What, then, are the advantages of roosting with somebody else, as opposed to singly and snugly? For long-tailed tits, and also goldcrests and wrens, it is all about huddling in physical contact and thereby reducing heat loss from a tiny body on a cold winter's night. However, for birds such as jackdaws or woodpigeons, which do not huddle, there must be a different explanation. In a few cases it might be because there is nowhere else to go, but usually the roots of roosting rules lie elsewhere.

The most popular hypothesis is that roosting communally is safer, for a couple of reasons. For one thing, being in a large number reduces your percentage chance of being killed by a predator. If you are attacked in a flock of 10 birds, you have a 10 per cent chance of being targeted; if you are attacked in a flock of 100 birds, this reduces to 1 per cent;

and so on, until the chance is infinitesimal. Of course, in practice, the risk of predation is much higher on the periphery of the flock, and other effects may play a part. The roost is never a single entity, either, but subdivides, reducing the value of the so-called 'dilution effect'.

Roosting communally may be safer for other reasons. Predators might simply be dazzled and confused by the mass of birds, making them less effective at hunting. Predators might also have a much-reduced chance of catching prey by surprise since there are so many eyes to spot them. And finally, roosts might be formed deliberately in areas where it is difficult for peregrines, sparrowhawks and merlins to operate.

Set against these hypotheses is the reality that large roosts of birds inevitably attract attention. Rare is the roost without a suite of hunters ready to attack, and when you visit a large murmuration of starlings you almost always see raptors, too. At night, owls might also view large gatherings as a tempting target.

One of the more interesting and left-field ideas that has come to prominence relatively recently is that a roost acts as an information exchange. This doesn't explain why birds join large roosts, as opposed to small ones, but there is some evidence that a bird's activities the following day may be influenced by other birds. The idea isn't that birds get together and have a good chat about how their day went and what they ate, but that birds can gather this kind of information by monitoring the condition of their flock mates. As we saw in chapter 2, birds are in some ways open books to each other, and it is easy to observe another's state of health. The theory goes that, if a bird that needs a good day's feeding finds another individual that has obviously done well recently, it could follow it out of the roost and to that individual's productive feeding grounds. There is some evidence that young swallows might do this. Adult and young swallows occupy different positions in their reedbed roosts, and so it is easy for the young to recognise the adults and leave at the same time in the morning.

The information exchange hypothesis does have some support in that, within every roost, there is a hierarchy, particularly of adults, which are dominant over the subordinate first-years. Recently, this arrangement was discovered in starlings spending the night on Brighton Pier, courtesy of infrared cameras set up by the BBC for their *Autumnwatch* programme. They found that the adults roosted at the top of the struts holding up the pier, in sites out of the wind. The lower rungs were occupied by first-years, with female first-years furthest down. The dominant adults, the lords of the rungs, slept by themselves, but the younger birds huddled together in bodily contact. As soon as the birds began the process of finding places to sit, there were loud calls and intense bickering, and there were outbreaks of aggression all night long. It doesn't sound very restful, does it?

Speaking of having a restless night, a delightful study of rooks speaks volumes about the harsh reality of life in a hierarchy. The researchers colour-marked rooks and studied them when they went to roost in some tall trees in the evening. They found that the precise location of a dominant or subordinate bird depended on the prevailing meteorological conditions at the time of going to bed. When it is cold, and particularly when there is a wind blowing, the dominant rooks quite naturally gravitate to the middle strata of the tree, where they are less exposed to the unpleasant conditions. When it was still, however, with little or no wind, the scientists were surprised to find that the dominant adults changed their perch and now occupied the top canopy at night. At first, they couldn't figure this out, but then they looked at the general condition of the subordinates at dawn. Some of them were spattered with bird poo and the penny dropped – and it wasn't the only thing that dropped. The dominant birds might have chosen the top perches in order to keep themselves from being peppered with excreta.

Anyway, back to starlings; if you have seen these birds at their roost and think that their arrival at the sleeping quarters is extraordinary, you haven't heard about their departure in the morning. It isn't as exciting to watch, but it is arguably even more remarkable than the dramatic evening show. Very few people have ever witnessed it and nobody has yet explained it. Starlings ought to be as famous for their morning routine as their night-time routine – you might scarcely believe what you read.

As the light grows in the morning, the members of the starling roost stir and, rather than chattering, they begin to sing. There is a crescendo, then the birds fall silent, for reasons that are not clear. The birds sing again, the noise builds up and, again without warning, there is another moment of quiet. After this happens a few times the birds get louder and louder until, on this occasion, the sudden silence is accompanied by a general flying around the roost, not in any clear direction. This is repeated until, with yet another dramatic shutting up, some birds lift off, fly in a circle and finally depart the roost in all directions. Having seen the first departure, the remaining birds once again build up their singing, go quiet and the next set of birds lifts away and departs, again without a common direction. These crescendos, silences and lift-offs continue until all the starlings have gone off for the day.

The astonishing thing is that it is staggered and there is a reasonably consistent time interval between lift off and departures – they occur about once every three minutes. Nobody knows why they should adopt such a time interval, but perhaps they need to leave space for each departure to find its direction of travel. Different birds leave at different times, as if setting off at a prearranged time, like a human timetable at a station, presumably in an order that is understood by everybody in the flock. As far as the scientists are concerned, nobody knows what the order means. Is there an advantage in leaving first – and if so, do the birds at the top of the hierarchies ensure that they are at the head of the mass? It would be logical to assume that the dominant birds depart first in order to get to the best feeding grounds before the others. But then again, why should subordinates let them? You can probably see now just how extraordinary and puzzling this behaviour by a common British bird happens to be.

While starlings remain quite mysterious, their roosting habits are at least described and well known. There are many British songbirds that we know very little about during the hours of darkness, and some that are profoundly mysterious. At the time of writing, for example, the usual roosting habits of as familiar a bird as the house martin are unknown. These birds are best-known for their mud nests, which were once a fixture under the eaves of many a house, both in the country and in more built-up areas, but are now retreating fast into notable nature sites. The birds nest in colonies and during the daytime they are usually aloft catching flying insects in the summer skies. They tend to fly in short arcs, gliding one way and then another, interspersed with quick bursts of their wings.

In the breeding season, house martins sleep in their nests. They apparently also have the delightful habit of frequently using unused nests during their migration, even if they have not nested anywhere near – a sort of house martin Airbnb. However, it seems that for much of their lives, including their stay in their winter quarters, they don't use nests for roosting, not even borrowed ones. They don't seem to use trees, and they certainly don't use reedbeds, as their close relatives swallows do. So where do they go at night? Nobody knows.

One possible answer is that they might spend the night aloft, well above ground, as swifts do. Perhaps, but when you see a house martin flying it doesn't seem to have the same aerodynamic skill that a swift has. One day we will know, when scientists have made data loggers small enough to fit on these birds.

CHAPTER SEVEN

Secrets and Robins

What can you say about the robin? No British bird has worked its way into the affections of the public quite like it. In 2015 it was voted as Britain's National Bird in an online poll, gathering 75,000 votes out of a total of 225,000 and trouncing everything else – it received nearly three times as many votes as its nearest rival, the barn owl. Much the same thing happened back in 1961 when *The Times* newspaper carried out a vote for the country's favourite bird and the robin was the top bird then, too. Neither the 1961 vote nor the 2015 poll is apparently yet fully ratified and so the UK does not yet have an 'official' National Bird, as do other countries as varied as the USA and Guatemala. Those in government responsible for upholding the recognition are mysteriously sitting on their hands. Perhaps it isn't easy to find the correct civil servant to sort this out?

Some of the headlines recording the robin's (entirely predictable) victory are telling. Take this one from *Country Living* magazine: 'Feisty red-breasted robin crowned Britain's National Bird', and this curveball from *The Times*: 'Expert sees red as robin poised to win bird poll'. The robin's reputation is not whiter than white. The first quotation refers to robins as feisty, which is a gentle *Country Living* way of saying violent. And the expert quoted, the organiser of the National Bird poll, David Lindo, reportedly denounced robins as overly aggressive and would have preferred them not to win. Not every headline was astute. The *Express* headline was 'Why we love red breasts'.

What better bird could there be to demonstrate the dark, realistic side of bird behaviour? One moment the robin is the fluffy, orange-breasted ball of feathers that delights us by its tameness; the next moment it launches into an unrestrained attack on a neighbour. With two default settings – 'cute' and 'seething' – the robin is the quintessential lovable rogue as far as humanity is concerned. But what is the truth about Britain's National Bird?

I went in search of the core of robins, to examine what life might truly be like in their world. I have watched them myself over the years; I checked out the literature and I sought out experts. I considered what people thought about them and I wondered what aspects of their biology might be surprising. It turns out that their lives are an open book of research. The most important aspect of studying robins was discovering just how complicated, as well as dark, the lives of birds can be.

So, what of the robin's reputation for being feisty and belligerent – is this true, and is the robin more violent than other birds? The witness for the prosecution will assert that robins are routinely spotted being aggressive and that they sometimes kill other members of their own species. In one study conducted in a dense robin population in Cambridge, a truly extraordinary figure of 10 per cent of all adult males (the sample size was 98 birds) died as a result of injuries inflicted during conflicts over territory; the figure for females was only 3 per cent. So the answer to the colloquial question 'who killed cock robin?' was cock robin.

This figure should be mitigated by the fact that, once neighbours settle down for more than a few days, serious fighting is very rare. The same thing applies to singing birds of many species. Nevertheless, for some days during territorial formation, robin society can be shockingly violent.

But are robins more aggressive than other birds? There isn't much data available for most songbirds, but the overall picture suggests that robins are unusual. Blackbirds are also extremely territorial, both in winter and during the establishment of territories in early spring. Fights are fairly common, but the impression is that they are less common than physical confrontations in robins. Blackbirds do sometimes kill each other, but it is rare. On the balance of probability, it seems that robins are more likely to kill their own species than other songbirds are.

Although the most violent fights between robins are known to happen in the early spring, I personally have witnessed the worst confrontations in the autumn. Perhaps it is because robins are so conspicuous then and you are more likely to catch them in the act. In September just about every other bird of the fields and hedgerows is silent, and the only real songsters are robins – with the odd exclamatory burst from a wren. The leaves of the woodland trees turn red, the sorrels and other herbs become crimson, and the bushes are full of scarlet berries. You might say that it is the right time to see red.

One robin spat has stuck in my memory because the battlefield was so benign. It occurred in what was then my favourite glade in the world, a quiet corner of the Isabella Plantation in Richmond Park, a Victorian woodland garden festooned with rhododendrons and azaleas and heathers under middle-aged oaks, a place where society is expected to be polite. In this genteel environment, two robins were singing their autumnal song, a less full-blooded version than the spring one, with longer, more muttering phrases giving the utterance a contemplative, wistful air. Somewhere in the narrative there must have been the robin equivalent of an insult, or perhaps an expletive. Or perhaps one robin turned and saw the orange breast of its rival glowing in the autumn afternoon sunshine? Whatever the cause, tensions boiled over before the metaphorical gas had been turned on, and the robins were immediately into a barroom-style brawl. The two birds were physically locked in combat, and their songs had turned into a strange, strangled squealing. For no more than a few seconds the birds violently pecked at each other as their feet tangled on a low perch – I have read that fighting birds often fall to the ground, still struggling and rolling, as the cowboys always used to do in western movies – and then they parted. I can only assume one backed down, because sustained combat is unlikely to have a happy ending. As it was, they both quickly resorted to a song duel of moderate vehemence, with presumably one gloating and one saving face.

What I witnessed was not the most spectacular robin fight you could see, with feathers flying and claws drawing blood. It was an everyday skirmish between two neighbours who were probably well matched. But what it did demonstrate was how quickly tempers can escalate and how much damage a real fight can do very quickly. The truth is that a serious confrontation is not something any garden bird enthusiast would want to see. Fights to the death are quick and brutal. We might assume that there would be much toil and struggle, and perhaps a degree of strategy and ebb and flow of fortunes. But, just as war movies exaggerate the drama of combat, so our imagination – and wildlife programmes – might exaggerate fights between animals. A deathly serious robin fight is quick and ugly and ends up with a lifeless ball of feathers.

It is easy to understand why robins should be aggressive in the spring – all birds are, however well they hide it. The stakes could not be higher for songbirds contemplating what might be the only breeding attempt of their lives. Failure is widespread and resists the birds' most profound urge, to reproduce. In the autumn, however, reproduction is not on the table. Indeed, your average bird in autumn has entered what is known as a 'refractory period' when the reproductive parts of its body all but shut down – the gonads shrivel and the bird is hormonally neutered. So disputes are not directly about reproduction. Why should they be so violent?

The simple answer is that, in contrast to the majority of British songbirds, robins hold a territory during the winter as well as the summer. While birds such as great tits and chaffinches roam around in flocks, robins spread out and defend the patch of ground where they spend the day feeding. As it happens, other songbirds also hold non-breeding territories but are either more flexible about their arrangements (pied wagtails either hold territories or join a flock), or are simply less obvious than robins about their territorial defence. One bird that is exceedingly violent in autumn and midwinter is the mistle thrush, but it isn't as common and well known and its aggression is overlooked. Tell that to a song thrush, though. Mistle thrushes latch on to berry-bearing trees with long-lasting fruit crops. These they requisition as a personal food store and will drive any rival berry-foraging bird away, be it another mistle thrush or another species. They will defend these food stores with up-front aggression and noise, but they don't have pretty red breasts, they are easily confused with other thrushes, and people don't see what they are doing.

In common with the mistle thrush, the robin holds a territory to protect its food supplies in the winter, but its diet is quite different. Although it eats a few small berries and seeds, it mainly relies on invertebrates, just as it does in the summer. And it is the way it procures this food that perhaps makes the robin the highly strung bird that it is. At the same time, incidentally, its foraging is the origin of the robin's celebrated tameness.

The robin might not conform to what most of us take for the term, but it is actually a 'sit-and-wait' predator, or at least, a part-time one. That means that when foraging it finds a perch overlooking a patch of ground where the activity of invertebrates is likely, and then simply scans for telltale movements, regularly changing perch if the cupboard is bare. Once it has noticed something, it then swoops down to the ground to snap it up. The robin's large eyes help it to detect small movements even in shaded ground and in low light intensity, and individuals have been known to spot action from as much as 20m away.

This feeding method is quite specialised. In fact, there isn't any other common garden songbird that practises it, although the closely related stonechat, of scrub and gorse, is an open-country equivalent. Robins do forage in other ways, including hopping on the ground with the intention of picking up whatever they might disturb, but sit-and-wait (or 'perch and pounce') sets them apart from their competitors. While many a garden species feeds in the leafy canopy or picks around leaves in shrubs, the robin is a low-level forager.

It isn't difficult, though, to see how sitting and waiting can prove to be a headache for robins. For one thing, in very cold conditions not many insects move about, so the foragers have to switch to ground feeding in order to survive. The method doesn't sound much fun in the wind and rain, either, when a gust could whisk the leaves of your precious leaf litter into annoying dust-devils and the raindrops keep the invertebrates wedded to the soil. And using an exposed perch could be chilly. One of the definitive images of robins, one seen on countless Christmas cards, is the ball of fluff surrounded by attractive snowdrifts. Robins routinely fluff their plumage when it is cold; the action of erecting their feathers increases the amount of insulating air trapped in the plumage, enabling the birds to better keep the heat in. It is possible that more active foragers don't need to do this so often.

The biggest bugbear of sitting and waiting, though, could be disturbance through competition. Imagine that you are having a difficult day, and the invertebrate ground activity is low. Then a beetle crawls into view, a significant meal. You are about to fly down to assuage your hunger when another robin, or perhaps a competitor of a different species, beats you to it by swooping in a fraction before you. It would be stressful. A similar situation could arise when two robins are feeding close by, and one disturbs the insect fauna by engaging in a bout of ground foraging before the other tries perch hunting. Proximity doesn't work.

What robins cannot abide, therefore, is disturbance. They don't tolerate it from any quarter. Any small bird caught trespassing is confronted by the robin's famed aggression, which usually has the effect of driving it off. Even wood mice, small rodents that pack a decent punch (I have handled them and they bite), have been ushered away by the small bird's red mist descending.

Once we understand how much a robin needs its privacy in order to survive, its volatility makes sense. Its territoriality makes sense. We see the violence in the context of the fight for survival that it actually is.

Some recent research has, incidentally, thrown a spanner into the well-ordered understanding of robin territoriality, meaning that much of what I have just written may eventually be undermined. The scientists discovered that a given robin population is not composed entirely of territorial birds at all, but there are some individuals that wander neighbourhoods and feed where they can when they can. In Spain in winter, at least, these birds do just as well as territorial birds, maintaining equally good health. So perhaps the true value of a territory lies in the ability of a territory holder to keep hiding places away from predators?

Whatever the true benefits are that accrue from keeping a territory, it is clear that almost every robin at least aspires to one, and that includes females as well as males. This introduces another quirk of robin behaviour, the tendency of females to sing. Robins are among the few British songbirds in which this happens, although their ground-feeding colleagues dunnocks do as well (see page 73). In the rest of the songbird community, song is almost overwhelmingly a male activity, to be used by females perhaps only in an emergency.

Female robins sing, though, and sing well. A recent study confirmed that female songs are slightly less complex than male songs and that both sexes can determine which gender is singing as they listen. Nevertheless, one detailed analysis showed that a female can sing at least a hundred song motifs, not far short of a male's repertoire, so it is very difficult for humans to tell the singers apart. Despite this innate ability, females only sing in the autumn. From the moulting period in July right through to the end of the year, the females are independent, but as soon as birds pair up, the male robin takes over all the territorial duties.

There is no more dominant sound in an autumn and winter garden and woodland than a chorus of robins, especially up until December when very few other birds are singing. One bird that most definitely would not be heard at the same time is the nightingale. Robins and nightingales are close relatives, but nightingales are transcontinental migrants that winter in Africa. Even when nightingales are here, from April to August, they don't sing for long, often becoming silent by June. Yet in recent years there have been more and more reports of nightingales singing in the dead of winter night, especially around Christmas time.

These reports are not wild fantasies fuelled by an excess of Christmas spirits – well, not all of them are, at any rate.

They simply nominate the wrong bird. The disembodied voice under the street lights is that of the robin. In the middle of the night at an emotional time, people exaggerate the quality of the song that they are hearing and attribute it to the finest songster they know. Yet the voice is the same as that heard in the morning and often at other times of the day in the autumn and winter; the ever-persistent robin. In the middle of the night there is no other song with which to confuse it, compare it and to demean it. The robin is the only act in town. In a concert with no Mozart to hear, we can appreciate Salieri.

It was once thought that nocturnal singing in robins was a response to artificial lights, because the phenomenon is heard in towns and cities far more than anywhere else. Those famous large eyes, so useful for detecting the change in light intensity and taking the lead in the dawn chorus, were thought to be playing tricks. We now know that, while light intensity may play a part, robins probably sing at night intentionally. A study in Sheffield found that there was a correlation between night-time singing and the ambient noise in the territory during the day. The noisier the location during the day, when most robins sing, the more likely a bird was to sing during the night. Everybody can appreciate how noisy towns and cities are in daylight, with traffic noise, aircraft noise, people talking and shouting, radios, dogs, sirens and even other birds. Against such a background, a number of birds worldwide are known to sing at night. They might do so to take direct advantage of quieter conditions, or they might do so to reduce the overall time they have to sing against the ambient noise. But the science suggests that they gain some advantage, set against the obvious costs of losing out on much-needed sleep.

There is one more aspect of robin singing that is easy to overlook, and that is this estimable bird's ability to mimic. As I write, robin mimicry is fresh in my mind, as I have been listening to them singing. Just recently I have heard robins making sounds that recall chaffinch song, blue tit calls, linnet songs, blackbird alarm rattles and even, a few days ago, the unmistakable trill of a little grebe, a sound like a manic version of the whinny of a horse. Some robins have been uttering little bursts of mimicry every few phrases. There can be no doubt about their source material – the copies are too good. But why they put in mimicry – or indeed, why any bird does – has not yet been satisfactorily explained.

There are hypotheses, of course, there always are. I find it particularly intriguing, however, that this one defeats an easy answer, because mimicry catches our attention, and seems obvious. Anyhow, one suggestion is that birds using mimicry are doing so not to sound like another bird, but simply catch fragments rather as we might catch a tune that we constantly hum. A second suggestion is that mimicry is done by mistake, at least in the case of birds that don't do it as part of every song, such as robins. A third suggestion is that mimicry is a way of increasing your repertoire, and thus to be more attractive to listening females. Mimicking a different species will ensure that you aren't using your own old material. This is analogous to putting quotations in an essay. It will still be your song, and will be sung within your own species' framework, but contains novel voices.

Despite the robin's ability to mimic and the extraordinary variety of its repertoire generally – one bird whose song was closely analysed turned out to have 275 different turns of phrase – you won't find many people who would say that it was a great songster. It is probably a little too high pitched and tinny for that. However, what the robin lacks in quality it makes up for in its year-round efforts. These days the robin's night singing could well make a popular bird more celebrated still.

But of all the robin's characteristics, none endears it to the public at large more than its remarkable tameness. I write this just after washing up breakfast in front of our kitchen window, enjoying the sight of a robin perched only a couple of metres away on a plant pot, giving a quick curtsey before flying over to feed its mate across the lawn. No other garden bird was close at hand, and it is typical for robins to be more confiding than other birds. They are easily tempted onto the hand with mealworms. They enter houses with little apparent fear of people. Not long ago a robin entered our house by the back door, flew upstairs (unbeknownst to us) and landed upon my computer keyboard. No doubt it was objecting to the articles I have written on robins over the years.

Robins simply don't have the same degree of repugnance towards us that most birds and other animals do, and for that, we are heartily grateful. The garden robin quickly becomes a favourite, and benefits as a result. We do not care if it is taking advantage of us. We realise that tameness is an unusual quality, especially in a small and vulnerable bird.

To embellish the relationship between British gardeners and their robins is an exclusiveness that is not entirely imagined. The British robin, which is a separate race (*Erithacus rubecula melophilus*) to the continental robin (*E. r. rubecula*), seems to have a more confiding nature than its European equivalent. Thus across the Channel, in France, and even more so to the south and east where it is a true forest bird, the robin declines to be as tame as in Britain. This can be at least explained by a general lack of historical persecution on these islands, as well as a long and distinguished history of home interest in wildlife and garden feeding. By contrast, bird catching has long been a widespread practice on the Continent and even persists to this day, shamefully, in parts of southern Europe. There is a reference to our own birds' tameness as early as the sixth century when a certain St Serf of Culross used to feed one that came to perch on his head or shoulder. Most references to robins in early French or German manuscripts are recipes. So perhaps British robins really are fearless; the only tables they will end up on are bird tables.

The tameness – indeed, impertinence – of robins comes into particularly sharp relief for many people when they are gardening. Your efforts to mow the lawn receive uncritical praise from the bird as it eagerly devours the invertebrates fatally stripped of their cover. Weeding, too, becomes less of a chore when the ants and beetles you unearth are snaffled as soon as your back is turned. What the robin probably most appreciates, though, is the tilling and turning of soil. The sit-and-wait predator barely sits and barely waits. The invertebrate fauna is unveiled and the robin runs rampant. This is a classic case of a biological relationship known as commensalism, in which one organism benefits from the actions of another organism, the latter of which isn't affected – except, of course, by enjoying warm feelings.

The robin's tendency to follow the actions of spade, hoe, dirty gloves and, yes, contented householder are probably ancient. Those who study robins have occasionally recorded them in wild situations following other animals, including pheasants and moles, and it doesn't stretch the imagination far to assume that, in the forest glades that are the robin's natural home, the birds might also follow large wild mammals such as deer and, especially, wild boar. In a robin's eyes any large, harmless mammal will do, and there are few more harmless mammals than gardeners. Shamelessly, the robin cashes in. It doesn't treat the gardener with a shred of affection; if there is a more active practitioner next door in the same robin's territory, it will defect. Having said that, robins might well be able to recognise human beings – there is plenty of anecdotal evidence for this – but then it is only as a patron might recognise a waiter at a restaurant.

Robins are so much a part of the winter garden that the last thing you might expect them to be is migratory. Yet they are, as a species. The Germans and Swedes have used robins in migration experiments, and to the latter, they are summer visitors, as swallows are to the UK. Of course, we have robins in Britain all year round, but if you go abroad robins are much more migratory.

Although I will look at bird migration in more detail in chapter 8, the robin's case is a good one for appreciating just how complicated bird movements can be. Robins in Britain could be categorised in three ways: stay-at-home, expats and transients. Firstly, the stay at home majority does not migrate at all. Most British robins spend their whole lives just a few kilometres from where they were born and are described as sedentary. Secondly, a very small minority of British robins leave our country in winter, as might a swallow, and then return in the spring. These don't normally go far, perhaps to Spain and Portugal, but ringing recoveries have proved that they do so. Thirdly, in the autumn between August and November and possibly again in the spring, Continental robins pass through Britain on their way to the south (north in spring). These birds are probably all from the north of their range in such places as Norway, Sweden and Finland; they will end up anywhere from southern France to North Africa.

I never realised that robins passed through the country until I read it recently, and then certain memories and curiosities from a career in birdwatching sprung back to life. I can remember arriving at a famous migration station in Norfolk and, while the bushes were hardly alive with flitting shapes, I can remember thinking thoughts such as 'Robins have had a good year this year'. Robins seemed to be on every bush, and of course, that doesn't normally happen, because they would be at each other's throats in such close proximity. On other occasions a whole patch of autumn scrubland would be alive with the 'ticking' calls of robins, suggesting a 'fall' of continental migrants. To me, there is no more evocative autumn sound than the staccato, clean-sounding 'tick' of a robin. I forever associate it with mists and to the waning sunshine and shiny red berries, of birds hopping next to the pink of autumn dock leaves and the scarlet of rosebay willowherb leaves. This to me is an image that comes with the rich smell of an autumn morning – and now I know it is partly made up from migratory robins.

So, not very far away from us in Europe, robins carry the seasons with them, leaving their northernmost breeding areas in the autumn and arriving back in the spring. As I mentioned, this has made them a popular subject for the study of migration. And robins were the guinea pigs in a famous experiment carried out in Frankfurt in the late 1960s by Wolfgang Wiltschko in which it was proved for the first time that songbirds could detect the earth's magnetic field. It was an elegant experiment with a decisive result, a scientific classic.

When a songbird is prepared for its migration, it betrays its inner urges by a restlessness that is palpable and also measurable. The pre-migration urge, or *Zugunruhe*, causes caged birds to flutter in the direction of migration in the evening when, in the wild, they would normally set off. This had been known for some time. So the scientists in Frankfurt developed a way of measuring the degree and direction of restlessness by using 'registration cages', circular units fitted with paper around the rim. The birds' feet were coated with ink so that, when they flew up to any part of the rim, their flight was duly registered. In this way the scientists could take a number of birds, each in a different cage, and get a measure of how often a bird fluttered to the south-west, the east, or whichever directions its inner drive would take it.

Previous experiments with robins showed that, on average, their urge was to fly south-west in the autumn, even indoors without any external navigational cues. Wiltschko and his colleagues wondered whether they could alter the robins' preference by subjecting them to an altered magnetic field, keeping every other factor constant. So they placed some large electrical Helmholtz Coils beneath the registration cages and using these, overrode and changed the magnetic north by 180 degrees, and then again 90 degrees. Every time they did so, the robins changed their direction of *Zugunruhe*, in line with their new magnetic environment. When the magnetic north was modified artificially, they oriented in the artificially modified southwesterly direction. This proved that birds use the earth's magnetic field lines (which run south to north between poles) in the absence of any other external cues.

Recently it has been shown that this could have been a weird experience for the robins concerned. There is some evidence that they might be able to 'see' magnetic fields – but more of that in the chapter on migration (see page 197). But if they can, imagine what they must have made of coping with a curious implement emitting strong magnetic radiation under their feet – psychedelic!

Earlier I said that there were three categories of movements of robins in Britain, but that isn't strictly true. There is a fourth one, and it is quite mysterious. David Harper, who has studied these birds for years and knows more about robins than is perhaps entirely healthy, tells about the 'million missing robins'. What happens is that, after the breeding season and usually after the moult, robins of both sexes and all ages take up territories. They might defend these for several weeks until, for unknown reasons, they seem to find staying put untenable. While most adult males do remain to fight it out between them, many juveniles and females up sticks and leave – they become migrants. They may depart in October or November, and some return as early as December, while others stay away until the spring. You might not think this is very remarkable, and perhaps it isn't – except that nobody knows where they go. Their movements aren't local, or they would be recaptured in ringing studies; there is no evidence to suggest that they go abroad, or even to the coast. They just disappear. Somewhere, in a big dark forest, they must be having parties with fairies. Or they go to Milton Keynes.

To a robin, being absent from its territory in the winter is probably not ideal for its future prospects, because these birds have another most unusual behavioural quirk – midwinter pairing. It isn't universal, and in migrant populations it is obviously impossible. But it so happens that, just at the time kind-hearted Britons are sending millions of cards with their picture on, the robins' sap is rising. In December, when day-to-day (and especially night-to-night) survival is far from assured, these birds effectively organise a future.

The cue for the change is when a male robin alters the way that it sings. Having for the last few months delivered its territorial song mainly from low down, half concealed (this probably protects it from predators), the male robin suddenly begins to vocalise from higher perches, in full sight, and usually from the centre of its territory rather than the periphery. This attracts nearby females, which come to prospect in the male's territory – very occasionally the male makes incursions into the female's territory instead. If both sexes are on the same page, they will indulge in a display called song-and-following. The male sings, the female chases, the male flies away, the female chases again, and so on until they have formed an alliance.

What happens next is equally intriguing. You might think that, now that the couple are an item, they would spend the next few months in twosome bliss. But however much bliss they might experience, they don't act as a twosome. They actually have very little to do with each other. The male goes back to singing its concealed territorial song as if nothing has happened. The two birds may not come into contact again for four months. However, in most cases, it does seem that the brief midwinter pairing ceremony is honoured.

Of all the facts that we know about the robin, one of the most surprising, and one that reflects the reality of its life most starkly, is its derisory life expectancy. With a mere

50 per cent survival rate, an adult robin can expect, on average, to live just one and a half more years. That is one, or possibly two breeding seasons in its life. Many small birds are similar. Of course, since this is an average figure, some birds do live very much longer, especially in gardens where they are fed by Brits. However, only about one in 10,000 robins live to be 10 years old.

If any aspect of robin life reflected the dark side of bird lives most completely, it would be this.

CHAPTER EIGHT

On the Move

Up until now, if you've got this far, you might have noticed that our journey into the dark heart of bird behaviour has centred on how a bird's life is more difficult, perilous and alien to us than we ever might have imagined. In previous chapters I have covered a catalogue of unexpected trials and tribulations concerning singing, feeding, protecting a territory, finding a mate and avoiding predators – so many, indeed, that you wonder whether any human could survive in a bird's world. If, after all this, you fancied a little light relief, this chapter is for you. That's because, when it comes to migration, the troubles faced by birds have, if anything, been exaggerated. Say it quietly, but it is true. Migration can be a bit of a doddle.

One of the first times that I noticed this was on the coast of Hampshire about 10 years ago, at a place called Needs

Ore Point. This is at the end of a peninsula which juts out a short way into the Beaulieu River estuary, not far to the west of Southampton Water. It was a sunny morning with a gentle easterly breeze, the sort of October day that reminds you of summer. As soon as I parked I could hear birds calling, and I disembarked excitedly, wondering whether birds had been on the move.

The calls, however, did not come from the bushes all around me; it was soon apparent that they were coming from the sky. A migration event had not just happened; it was happening now, at this moment. Normally in the early part of the autumn, you arrive at a coastal site hoping that you can locate unusual birds that have landed at dawn after migrating during the night. Unless weather conditions are poor, much of the migratory business can be completed shortly after first light. On this occasion, however, flocks of birds were still passing over during the morning.

It was one of my first experiences of what birdwatchers call 'visible migration'. This refers to the movements of birds during daylight hours, usually first thing, when they can in theory be seen in the act. Birds that begin their journeys at dawn, as opposed to sunset, include a suite of species such as finches, pipits, wagtails, swallows, buntings and larks. On the other hand, warblers, chats and flycatchers tend to migrate at night, a subject I'll return to later. The important point is that you can actually witness the journey taking place.

The overall impression I got on that day was one that has always stayed with me on subsequent observations, and this is how leisurely the process appears to be. As I watched from Needs Ore Point, every so often a charm of goldfinches would pass over, all cheery, tinkling calling and brilliant yellow wing-bars. They flew into the soft headwind, not very high – perhaps not much more than head height, and not very fast. The flock didn't hold together very well; every so often an individual would wander away from the body, only to reel itself in as if attached by elastic. They exhibited all the hurry and determination of a bunch of humans window shopping on their day off work. If goldfinches were ever tempted to take selfies, these apparently highly relaxed migrants would have done so.

They weren't the only ones that made the journey look easy, either. Small flocks of linnets passed by, just as low and leisurely as their fellow finches, making their more metallic calls and flashing white wing-bars rather than yellow. The odd skylark flew higher in the blue sky, but not any faster. Wagtails bobbed over, and so did small parties of meadow pipits. The pipits, which are related to wagtails and have slender bodies and quite long tails, but look like miniature thrushes, seemed to take their movements still less seriously than their counterparts. Some of these birds interrupted their flights just to circle around, even to touch down briefly, before moving on east over the water.

It would be easy to dismiss this stream of birds as residents, perhaps making the short hop across the Beaulieu Estuary in order to forage at some rich location. It is possible that the blue tit that flew high over my head, calling, and heading out to the crossing, was doing just that. And perhaps the crows and rooks were merely commuting and not migrating. But the number and variety of species told another story. The swallows were surely moving; by this date, 12 October, they were behind most of their colleagues. Redpolls and siskins, forest-dwelling finches, were appearing here in a place with hardly any trees. A grey wagtail, far from the running water it depends on, was presumably not from these parts, surely travelling, a genuine migrant. This was migration, alright, but not as many might recognise it.

The movements noted on autumn mornings between September and late November occur throughout the country every year. Owing to the fact that they take place in daylight, it isn't unusual to witness the birds actually finishing for the day. I have seen this myself on Hengistbury Head, in

Dorset. It was a similarly fresh, sunny morning to that at Needs Ore Point and, on the clifftop, you could see and hear the birds passing just above head height. There was clearly a heavy passage of meadow pipits. Small parties came through low, heading east, for several hours. And then some began to stop; rather than departing, they simply landed and started feeding, wandering over the grass sward. Pipits walk, rather than hop, and these individuals scampered to and fro, stopping to pick at the ground for a small morsel of food. They remained in place for the rest of the morning, showing no intention of moving on.

The autumn movements seen as visible migration raise a number of questions. The most obvious ones are the ones we always ask about migrating birds – where do they come from and where are they going? However, I am going to skip over these questions – they have different answers for each species, and sometimes for individuals within a species – in order to concentrate briefly on another one: how far do these birds fly in a morning? Once again the answer will vary from bird to bird, but you can work out an approximate answer by taking the typical speed of a swallow, for example, and multiplying by the flight time. A swallow flies, very conservatively, at 30km/h (18.64 mph). If a bird sets off at sunrise in mid-October on the south coast, which is about 7.30am, and then continues until about 11am (the time by which most visible migration has ceased), then its flight time is three and a half hours, meaning that the distance travelled is only just over 100km (60 miles). Doubtless the swallow, with its swept-back, pointed wings adapted for aerodynamic efficiency, would qualify as one of the faster flyers, so it seems doubtful that any of these morning migrants would cover any longer distance. And that really isn't very much.

What these movements show is that, despite the rightly celebrated feats of migrant birds, not everybody is in a hurry.

It is true that some small birds, such as sedge warblers, make exceptionally long flights without stopping. These sprites feed up on a special kind of insect (the plum-reed aphid) in northern France, and then apparently fly all the way to West Africa in a single flight of seventy-two hours (see page 181). There is no doubt at all that birds can perform remarkable feats of marathon flying, but it is equally true that the majority do not. To a bird, three and a half hours' flying is no more unusual than is three and a half hours' walking to us – it's peanuts. Most migration is undertaken, it would seem, in moderation.

And while a wagtail may not travel much further than from Scotland to southern England, and while a linnet may not move more than 50km (30 miles) from its breeding sites to its wintering sites, even those birds that undertake vast transcontinental journeys, often flying at night, will not necessarily take giant steps, especially not at first. As eminent migration researcher Peter Berthold says in his summary of a 'typical bird's migration' in the book *Bird Migration: A General Survey*: 'The first phases of active migration, like many subsequent nocturnal flights, will be of relatively short duration.' He continues: 'With the completion of juvenile moult and the increase of fat reserves, the migration stages become slightly longer and the resting periods shorter, finally leading to nocturnal migration every night. However, the migration stages are still confined to a few hours of flight every night, as the average migration speed towards the winter quarters is only about 50–75km (30–46 miles) per day.'

The conclusion is stark. Migration for most birds, in most instances, is not particularly taxing. And why should it be, especially in autumn when birds have completed breeding and are not in a rush, as they are in spring, to re-take the previous year's territory? Birds begin their migration considerably earlier than many people realise. A common migrant bird such as a willow warbler could leave its territory in mid-July, right in the middle of summer, and its first movements could be no further than a human's walking distance to the other side of the parish. The declining days of autumn in fact give birds ample time to reach the wintering areas. Some individuals will be forced to hurry by circumstances; not being in good condition, for example. The rest have no need to set off when they don't have to. They can wait if the weather is poor, and they can postpone a flight if they have had a bad day foraging for food.

And yet you might be forgiven for assuming that, the moment a swallow or other much-loved character sets off on its migration, it suddenly plunges itself into peril. The list of dangers it might face is oft-quoted: bad weather, predation, exhaustion, getting lost, starvation in the desert, and so on. Faced with such a daunting raft of hurdles, you might expect that any sane bird would simply fail to set off. But of course this is nonsense. Statistically, for a bird such as a swallow or a nightingale, the one sure way to come to grief is not to set off. Birds migrate for survival reasons. The plummeting supply of insects that marks the end of autumn and the beginning of winter is a warning that, for many insectivorous birds, the months ahead are not sustainable unless they move on.

One of the key points covered in this book is that birds spend much time in grave peril anyhow. Their lives are short, and problems face them wherever they are. Many of the above perils are daily perils: take bad weather, predation and starvation, for example. They are as likely to succumb to these on their British breeding grounds as they are in transit elsewhere. And if they prepare well for their migration, by feeding up, getting in shape and not rushing, they can eliminate exhaustion. There hasn't been much work done by scientists on birds getting individually lost, although we know it does happen. Birds doubtless make mistakes in their navigation and orientation.

The popular understanding of migration is also clouded by an over-emphasis on long-distance flights of popular and photogenic birds such as swallows. These birds come north to us in the spring to breed and leave in the autumn for warmer climes. About 25 common songbirds fall into this category and, apart from swallows, they include much loved species such as nightingales, house martins, willow warblers and spotted flycatchers. It is hardly surprising that people long for the day in spring when they are reunited with these birds after the long, dark months of winter. Equally, we say goodbye to them in autumn with a heavy heart. But there is a great deal more to the story of migration than the arrival and departure of so-called 'summer visitors'. As we have seen, much inbound migration takes place in the autumn, too, from September to November, although people tend to notice and care about it far less. They see swallows perched on wires in the early autumn, looking a little like musical notes on a stave, the signature tune of the changing seasons; they might sigh about the impending departures, and then shut up shop in their minds. Unless they live in regions where the call of wild geese or wildfowl signals the beginning of winter, they will perhaps be unaware that the biggest migration of all, in the autumn, doesn't drain our country of birds, but actually swells it.

A Briton, in particular, might be surprised to hear that our very own damp and chilly island is a major winter destination of choice for millions of bird migrants. To these seasonal visitors, we are 'warmer climes'. Britain has a milder climate than most of the continent, and very much warmer conditions than Arctic Fennoscandia, which is something of a short-haul flight away. Many people are aware of geese and waders coming south to Britain, but fewer are aware that individuals of many familiar species also use our Sceptred Isle. For example, in October we are 'invaded' by several million chaffinches, perhaps a million blackbirds, and other songbirds including goldcrests, pied wagtails, skylarks and song thrushes. These birds have come to stay for the winter. Their arrival is masked by the fact that we don't recognise them as incoming migrants. We think of them as resident birds, and so they are. They are both residents and migrants.

A good example of a bird that comes to Britain for the winter in large numbers is the starling. There are about 2 million pairs breeding in Britain (they have declined by about 50 per cent between 1995 and 2010), but in the winter there are probably at least 2 million immigrants (the last estimate of 37 million was in the 1960s, since when the population has declined). That means that they are common and familiar at all times of year, but they happen to be more abundant between October and March. Not many people would notice the difference, though, so the starling is not generally regarded as a migrant bird. This is not the case in Finland, where all the starlings go south in the autumn, and where the first sighting of a starling arouses as much delight as the first swallow of spring does here. However, we cannot distinguish a Finnish starling from a Finsbury Park starling, and so the starling's migration is overlooked.

Our wintering starlings come from many breeding areas. Some simply cross the North Sea or the Channel from the Low Countries, often arriving as early as September; another gang comes from Germany and Scandinavia, arriving in October; and yet another arrival is from Poland and Russia, often in November. In some areas they gather in their thousands. We should perhaps be proud that they come here; why not?

It so happens that starlings do make themselves conspicuous. They are celebrated for gathering together in enormous swirling flocks, which paint pictures in the sky above the roost sites, creating a unique canvas each evening. Starling roosts are hard to avoid in wildlife films, but less easy to experience first-hand – or at least, the enormous multitudes of starlings are comparatively rare. Being below a swarm of starlings has the feel of being underneath a thundercloud, with its own extraordinary atmosphere – created by fast-moving living bodies – imbuing the watcher with a similar sense of awe. When swirling the birds are silent except for their wingbeats, and the sheer spectacle silences most human witnesses as well.

The giant roosts of starlings are a winter phenomenon; from spring through to autumn the resident population does roost together, but their numbers are never impressive until winter sets in. And that is, of course, because of immigrants. The millions-strong starling show is an imported natural wonder. If Britain was not the winter refuge for Continental birds, it wouldn't happen.

In fact, the cultural emphasis on popular species arriving in spring and departing in autumn masks the reality of migration as a year-round phenomenon. It is true that bird movements are greatest between March to May and between August and October, but songbirds move in every month of the year. I want to examine two aspects of this.

Not long ago, I found myself on a degraded but biologically diverse mountain in Malawi, a small republic in Africa named after the lake which covers a third of its area. The Zomba Plateau is, of course, far away from southern England, but the geographical distance is at least equalled, if not eclipsed by its sheer different-ness. Visiting south-central Africa, especially in November, is like the metaphorical switching of station from low-key piano sonata to a loud blast of high-energy rap. Southern Malawi is dry, dusty, agrarian, astoundingly populous, joyful, personal, warm and economically poor. The wildlife is amazing where it is protected, but here near Zomba, little forest is left. The point, though, is that there are so few ecological and climatic parallels with the UK that you have no real reference points for comparison. And yet here, rubbing shoulders with exotic birds such as Livingstone's turaco, red-faced crimsonwing, white-starred robin and forest double-collared sunbird was the humble garden warbler. I heard it singing among the tropical-sounding voices, its stream of babbling subtext emanating from a bush that was thick and thorny and could have been itself translocated from a slope of British downland. Later on, not far away, I watched willow warblers on acacia trees, literally feeding a few metres above vervet monkeys and groups of grumpy hippopotami. It might be overdoing the point, but there is nothing like seeing European breeding birds in their wintering context to understand that migration doesn't finish when a bird has left our coastlines. That garden warbler probably spent the months of August to November migrating, and will be ready to start back as early as February. We should therefore not forget that our own breeding birds are travelling for much of the year.

Even if we remain in Britain, or stay in one spot in Britain, similarly there is something going on every month, every week and, well, more or less every day, if not quite. And that's even the case among songbirds, though it is easier to observe with much larger birds as gulls and ducks. In particular, the autumn migration starts much earlier in the summer than you might expect. In the case of those same willow warblers mentioned just now in the savanna of Africa, the adults have usually completed breeding in late June, and their juveniles even then are wandering around the neighbourhood, in a form of migration known as dispersal – more of that in a moment. The juveniles begin their formal, southward-oriented migration in the second half of July, and so do those of many other species such as common whitethroats. By August the adults are on their way, too, meaning that the whole population is on the move during the human school holidays. Several of our famous migrants, including nightingales and swifts, have all gone by the end of August.

In recent years the BTO has been radio-tracking cuckoos, and their survey has redefined the migration season for this particular bird. Remarkably for an undoubted 'summer bird', most adults leave the country before the end of June. In fact, one individual was recorded leaving for Africa on 3 June, having arrived only in late April!

The summer movements of juvenile birds are an interesting phenomenon that most of us don't credit as migration. It depends how you define migration, but if you broadly accept that it is a large-scale movement of living creatures from one place to another, it just about fits the bill. What happens – and this is something that you can easily see in your garden, or local woodland and scrub throughout July – is that juvenile birds leave the territories of their parents, a physical emigration that might take them just a few kilometres. They usually team up with a gang of other birds in the same position and wander around, fairly aimlessly, in a mixed flock. After a week or two of this, they usually begin their autumn migration.

This so-called 'post-juvenile dispersal' is a phenomenon that everybody can enjoy in what is a quiet time for birdwatching as a whole. Spend enough time outside in a scrubby area or woodland edge and you are bound to see it – suddenly a bush is alive with birds of many kinds. You might have a family party of long-tailed tits (adults and parents) forming its core, but there are likely to be blue tits, great tits, coal tits and perhaps marsh tits joining, too. In many flocks there are juvenile nuthatches and treecreepers, goldcrests and, if you are fortunate, you might find warblers, too, such as blackcaps, whitethroats, willow warblers and chiffchaffs. A good flock may have fifty or more birds in it, each calling and collectively making enough of a rumpus to catch your attention. The birds will be on the move, flying from bush to bush quite quickly, so that you might need to go at a slow walking pace to keep up. On a sunny day, with the light shining on the turning leaves and with new, plump berries adorning the bushes, a large flock of dispersed juveniles – often with the ragged plumage of youth as they replace juvenile plumage with first-winter feathers, and often with a yellowish wash (the cheeks of blue, great and coal tits and the whole hue of chiffchaffs and willow warblers) makes for a delightful birding encounter.

There is one particularly unusual event that occurs in midsummer and could perhaps be described as a migration. It concerns that quintessential cricket-season bird, the swift, the supremely long-winged dark-coloured waif that flies around rooftops screaming. Swifts are specialised feeders on very small aerial invertebrates, including some terrestrial ones like baby spiders, that are whipped up with summer breezes and ride them for the purpose of dispersal. While there is plenty of food available when warm, still air wafts over the land, supplies of these powder-puff invertebrates drop sharply when it is cold, windy or rainy. It seems that the swift can tell when a cold front is approaching, and in certain circumstances it will sense that conditions are about to become unsustainable. You might think that it would retreat back into the nest and sit tight when this happens, but instead it simply flies away from the problem, making for clear skies. It might have young in the nest needing to be fed, but happily these are well adapted to cope with starvation for short periods of time, and can lower their own metabolic rate. Sometimes the adults fly hundreds of kilometres away from their breeding areas, even intruding into the airspace of the Continent – they literally fly around the circulating air making up the cold front. They might also make a shortcut south-west to reach the warmest sector of depression front promptly. To keep themselves in good shape they must feed. These so-called 'escape movements' enable them to reach good feeding areas in times of hardship – and this in the middle of summer!

The great southward move known as the autumn migration fully occupies the months of August, September and October, and only tails off well into November. Before mentioning what comes after it, winter migration, I cannot resist mentioning a small bird that entirely bucks the trend of autumn by flying northward instead of south. Very few people have ever heard of it, but the water pipit is a perky little bird that walks around the edges of lowland freshwater pools in southern England and blushes a deliciously pleasing peach colour in the spring. It is the only bird species that breeds in the high-altitude meadows of Central Europe, such as the Alps, Pyrenees and Carpathians, and then comes here. It is really an altitudinal migrant, simply descending from the mountaintops and wintering in the comparatively mild lowlands, but it is always intriguing to see it here while imagining its summer life, among the cowbells, flower-rich meadows, snowfields and ski-lifts at rarefied altitudes.

Migration does happen in winter, albeit at a much reduced level of traffic. The months of December to February usually see birds settled, and ready to do battle with the difficult conditions and with each other. That said, some British winter visitors are virtually nomadic, which means that they can move from place to place at any time, depending on food supplies. A good example of such migrants are the so-called 'winter thrushes', the redwing and fieldfare.

These lovely birds really do brighten up the winter. Well turned out and colourful, they are also lively, sociable and flighty. Their breeding heartland is Finland, Norway and Sweden, and it is from here that the majority come on the well-worn short hop-over across the North Sea. Their calls are familiar to every birdwatcher who has ever ventured into farmland fields and snowy hedgerows when the wind and cold are raw; and you can also hear them at night as they make long or short movements just above rooftop height. The redwing has a whispering 'tseep' call, and the fieldfare a gentle, slightly slurred 'sh-shack', and the two sounds contrast perfectly, one piercing and one clucking. They are the symphony of berry-clad hedgerows, ploughed fields and open woodland. When you hear the calls, you can see or you can visualise your breath vapourising in the keen air.

Both species – in common with many others in the autumn – feed mainly on berries for as long as they can if supplies hold, and it is this fruit-eating habit that ushers them from place to place as stocks reduce locally. It is also this, indeed, that forces many over here in the first place; if there is a heavy crop of berries in Scandinavia, many will delay their transfer to the United Kingdom until late November and December. Once they arrive they may make significant movements overland, but by Christmas their wanderlust is usually tempered; they won't make many movements of more than 10–20km (6–12 miles) until March, when the urge to move north is too much.

Just occasionally, winter conditions become severe in Britain, with plunging temperatures, heavy snow and frost. Although routine in Scotland, this is still very unusual over most of the rest of the country, and while it might take us by surprise, it is much more of a shock to our birds, particularly if the spell of poor weather is sustained. Most birds can cope with a day of poor or zero feeding, but any longer than this can quickly turn into an emergency. Many birds succumb to snow and ice, but others enter what might be termed emergency migratory mode. They effectively up sticks and leave the area, temporarily or permanently, and become refugees. They are not prepared and one can assume they don't go willingly; they evacuate an area to save their lives. Whether or not it works depends on how quickly they can reach an area where they can feed again.

Quite a number of species engage in these midwinter 'escape movements', including redwings, fieldfares, song thrushes, blackbirds, skylarks, pipits and finches. Many aren't evacuating the snow or ice as such, but simply need to reach unfrozen soil, where they stand at least a chance of finding something edible. I have witnessed a few of these migrations, and they can be hard to watch. While many birds fly strongly, not all do, and you can usually see weak ones grounded, moving slowly and looking lost. It isn't the cold to which they succumb – feathers are usually a more than adequate buffer against cold – it is starvation. The fire goes out, slowly, and it isn't pretty. Observers of winter evacuations have witnessed birds simply dropping dead out of the sky.

Not all winter movements are so dramatic or torrid. In recent years, as scientists have become better at tracking birds and following their movements, more and more individuals – as opposed to members of a species – have been uncovered as winter nomads. Among their number are individual blue tits and chaffinches. While the main body of individuals of a species stays true to an area, a few are floaters, arriving at a garden feeder, for example, for a few days before inevitably moving on. They come into a new town like the lonesome cowboy in the generic western, never getting to know a personal flock or finding their place in a long-term hierarchy.

If you add these small-scale movements to the general migration story, you realise just how ubiquitous it is, and how it goes on all the time at different scales. Of course, now we know that insects migrate, as well as fish and mammals and all sorts. We shouldn't be surprised, but we can still be amazed.

And happily, in contrast to many of the aspects of bird behaviour described in this book, migration really can be described in all the ways it often is – incredible, amazing, uplifting – without reality soiling the brand. As we've seen, migration does not always carry the humongous risks often attributed to it, and can be seen as a positive response by remarkable creatures to the ecological pitfalls thrown at them. As such we can enjoy migration with a clear conscience.

How, though, can somebody best come to appreciate migration for themselves? Speaking for myself, I could recall memorable days of mass visible migration and 'falls' of small birds appearing overnight, but these are merely highlights, the extracts suitable for photo albums. Getting to grips with migration in a personal way, a way that brings home the sweet drudgery of bird movements, is different. I have managed this by becoming involved with a ringing scheme dedicated to monitoring the birds of a particular place.

Longham Lakes, near the overlooked town of Ferndown ('I think I've driven through it') in Dorset, is a perfect example of what is known to birdwatchers as a 'local patch'. Nobody would come here if they had a better offer; there are literally dozens of excellent birding locations a short distance away, many of which attract enticing rarities. Many of the town residents have not only failed to visit Longham Lakes, but haven't ever heard of it. Yet the two reservoirs next to the River Stour and their surrounding woods, farmland and scrub do attract enough birds to make regular visits worthwhile, so long as your expectations never become overheated. It is a good local walk, too, with pleasant countryside, but again, with dozens of other prettier locations well within reach.

It is this unprepossessing stage where I have had my best front-row seat to witness migration. Here a friend of mine, Roger Peart, has enrolled into a scheme to monitor the local birds and make Longham Lakes a 'Constant Effort Site', which involves visiting Longham twelve times throughout the summer to catch birds and affix small rings on their legs. He also rings in the spring and autumn to pick up any visiting birds passing through. It is laborious work. He must arrive on site before dawn, and make detailed measurements of any birds he catches. I occasionally join him, albeit only as a fair-weather observer, arriving at sociable hours. I am a bit-part observer, although I do put his results online.

Observing the ringing of birds never ceases to enthral me. For one thing, I am still not used to seeing birds in the hand, at such close quarters that you can almost feel their energy.

Birds are always smaller in the hand than you expect, and never quite resemble the image you get through binoculars. Indeed, it is perfectly possible not to recognise a bird species in the ringing net or in hand that you see all the time in the wild. On these intimate occasions, too, you appreciate how every bird is an individual. Some will look smarter than others, some will be shabby. Some will object to being handled, while others simply let you get on with it, as if they were sitting quietly in a dentist's chair. Many try to bite you and scratch you, and you remember that they are all survivors in a dangerous world.

Perhaps the most enthralling part of ringing, however, is the net round. When engaged in a Constant Effort session, most ringers will set up a number of nets, all carefully placed in spaces across which birds are likely to fly, for example between two bushes. The nature of the terrain usually means that the nets are relatively hidden and you cannot see them all at once. Every half hour or so, the time allowed between net rounds, you therefore embark on a mystery tour. You have no idea how many birds you have caught, if indeed there are any, and you will not know what species they might be. It could be anything. Nets often catch surprises, birds you had no idea were present. The nets catch noisy species and quiet species, obvious species and the most secretive. Nets catch common and rare species, whatever happens to be around – and that depends on a bewildering range of factors. Net rounds are entirely unpredictable.

In some ways, a net round can be compared to opening the post, in the sense that you don't know the contents of the post until you receive it. And in a still more apt metaphor, the post could be coming from anywhere, far away or close at hand. Such is the reach of our migrant birds that individuals at any location could be arriving from a distant destination, or from the field next door. Even on the same day, two individuals of very similar or even the same species could have arrived from completely different destinations and have quite divergent goals. For example, at Longham Lakes Roger has ringed two warblers, the whitethroat and lesser whitethroat, on the same September day. However, British common whitethroats migrate west and migrate over Spain to winter in West Africa. Lesser whitethroats, on the other hand, head south-east to Italy, the Balkans or even the Holy Land, and they end their journey in East Africa, more than 3,000km (1,864 miles) away from the common whitethroat's destination. They go from bush mates in Britain to very different parts of the African bush.

Another intriguing aspect of ringing is that the most memorable sessions might not be memorable until much later on, sometimes months or years afterwards. Not long ago I arrived very late on an October day to find Roger ringing a blackcap, a small songbird with an eponymous black skullcap (brown in the female). Bearing in mind that he ringed almost a hundred in the course of that year, it was hardly a moment for trumpets and drum rolls. But a few months later that same blackcap found its way into another ringer's net, but this time in southern Portugal, 1,598km (993 miles) away. When we did eventually hear the news, it was tremendously exciting, proof that this small bird on our own patch was migrating a substantial distance. Indeed, it was I who 'got the last touch' of the blackcap in Dorset, releasing it from my hand back into the wild.

Such experiences make migration personal. They also make migration seem what it really is, routine and organic. On other net rounds we have caught a small bird called a reed bunting. This is a species that few would ever call their favourite; it is sparrow-like, has a rather unimpressive song, is often rather secretive and doesn't seem to 'do much'. Our ringing sessions prove, however, that a number of our breeding reed buntings make a journey to the coast to spend the winter. The distance of this migration? It's just 10km (6 miles)! On the morning of our reed bunting's departure south we are not going to wave it off with dread in our hearts hoping for the best, are we? This is the small change, the nitty-gritty of migration, and in some ways the heart of the matter. Our reed buntings find our patch of inland Dorset untenable in winter (we have few if any records at Longham in winter, despite a breeding population of at least ten pairs), so they disappear to the nearest piece of habitat that will support them. It is gloriously unremarkable.

Birds are practical creatures. They won't migrate a long way if it isn't necessary. A wide range of British birds simply decamp to the Mediterranean for the winter, just as some people do.

Of course, while migration might be very natural for birds, and not necessarily a huge burden, that does not mean that we should ignore the many remarkable feats carried out by some songbirds. For whatever reason, some species fly enormous distances, sometimes in lengthy spells that make us wonder at their feats of endurance (and at their navigation, too, because endurance without direction would have disastrous consequences). The great flights are the ones that tend to get most people excited about migration.

The sheer distances routinely covered by common British birds are truly extraordinary. You might have heard that, in former times people suggested that swallows and other migrants could have overwintered underwater in ponds. We scoff at such a suggestion these days, but actually it is a perfectly plausible idea – male frogs may do exactly this, so why not birds? If you had suggested to a pre-nineteenth century naturalist that swallows fly all the way to southern Africa for the winter, 12,000km (7456 miles) away, they might well have considered the idea preposterous. In the days before cars and trains, when the fastest mode of travel was horseback, people would have found it hard to believe that tiny birds could make transcontinental journeys under their own steam.

Yet as we now know, a fair proportion of our birds do just that. A significant number of British species of birds make a regular journey that takes them further from our islands than many people ever go. Those people that unfortunately cannot afford to travel, or who go on holiday only to the Mediterranean, or to the Canary Islands or continental Europe, don't range further in their whole lives than the swallows, flycatchers, willow warblers and swifts that share their local area. We are often out-travelled by creatures that fit into the palm of our hands.

Before looking at some details of migratory flights, I want to examine a line of reasoning suggested by a number of authors that really does turn our idea of bird migration on its head. How might we view transcontinental bird movements if we consider Europe (or northern Asia or northern North America) as a bird's destination rather than its origin? Normally we don't do this, because species like willow warblers breed here and their first ever journey after hatching is south towards Africa. However, if we look at a couple of different migrants, the cuckoo and the swift, we get a different perspective.

Recent radio-tracking studies on the cuckoo by the BTO (you can follow them online) have shown that the birds arrive in April or May and have all departed for Africa by the end of June. One particular tracked bird was found to spend 38 per cent of its life migrating and 47 per cent of its life in Africa. That leaves just 15 per cent of its existence spent on British soil.

The swift is another bird that does not remain here for long. Rarely seen before the very end of April or first week of May, swifts remain for the first part of the summer, but few linger beyond the end of August. They spend a maximum of four months in our airspace, and some birds, those unable to breed, are airborne for the full duration of that stay. Several other species, including nightingales and pied flycatchers, also arrive in April and leave by the end of August. Their stays are brief and businesslike, whereas in

Africa their time is far more leisurely, unencumbered by the demands of breeding.

On the basis of this, can we call a cuckoo, for example, a British bird? Well, we can in terms of nationality: it would have been hatched in this country and so, in all likelihood would its parents. Their passports would say 'British'. However, for tax purposes, the cuckoo would perhaps be a non-dom. It spends little time in Britain, and far more in Africa.

Some scientists postulate that many of our transcontinental travellers are actually African birds at heart, and not 'ours' at all. During the last series of Ice Ages, Arctic-like conditions on the European mainland would repeatedly have squeezed what songbirds there were back towards North Africa. Scientists have long suggested that the great transcontinental journeys arose originally from short hops that the birds made from their African heartlands in summer, first dipping their toes on the European mainland and later, as the ice retreated for the last time, gradually creeping ever further north summer by summer. Each migration north would have been followed by a return south to avoid the Ice Age winter, but year on year the journey could have got longer. If we accept this simplified version of the evolution of migration, we would have to accept that Africa is a swallow's and a cuckoo's ancestral 'home'. Coming to Britain, an outpost off the European mainland, would simply have been an opportunistic attempt to breed in a niche that wasn't filled and, better still, to take advantage of the long summers of the newly unfrozen lands. Indeed, you could argue that some summer migrants only come here because the long hours of daylight allow them to squeeze in more productivity (often second broods) than would be possible in tropical Africa. After breeding these birds revert to type, becoming African again after their brief sojourn.

That is certainly a different way of looking at our migrants. Their brief visits to Britain and Europe are not so dissimilar to visits by seabirds to their breeding cliffs each summer. Birds such as puffins and guillemots breed on dry land, but that doesn't mean we think they are landbirds. We know they are seabirds. By the same token, are not cuckoos African birds?

If migration did evolve as described above, it might explain why some birds these days make such awesome migratory journeys. A classic example is that of the Greenland subspecies of a widespread British breeding bird, the northern wheatear. This bird has a difficult journey across dangerous terrain to take advantage of the long and lustrous northern hemisphere days. Its maximum migratory journey of perhaps 15,000km (9320 miles) is thought to be the longest undertaken by any songbird. What is truly remarkable, though, is that some include a sea crossing between Greenland and the European mainland, amounting to 2,500km (1,553 miles) non-stop. Obviously, it cannot stop anywhere on those wild seas. I myself have seen this bird in the Azores, no fewer than 1,500km (932 miles) to the west of Portugal.

Another particularly intriguing case concerns our old friend the sedge warbler, the same bird that has proved such a fruitful subject to study song and its meaning. As mentioned above (page 163), the evidence from ringing suggests that these tiny birds might undertake a single flight from southern England or France right across Spain or Portugal, over North Africa and the Sahara and right down to West Africa below 17 degrees south. It isn't actually the ringing recoveries that suggest this, but the lack of them. In contrast to the reed warbler, a relative that is often found in similar habitats in the breeding season, sedge warblers are hardly ever trapped or recovered in the Iberian Peninsula, or in North Africa. Although it is hard to prove a negative, the fact that many reed warblers have been found on migration around the western Mediterranean does suggest that their sedge warbler colleagues are doing something very different. The evidence suggests a non-stop flight of about 3,000km (1,864 miles), while reed warblers make a similar journey in short, almost leisurely hops.

There is a truly intriguing subplot to the sedge warbler's autumn migration, and this concerns the bird's diet as it prepares for its departure. Sedge warblers tend to breed on the margins of wetlands; their nests are in bushes above dry ground, as opposed to reed warbler nests, which are typically in reedbeds above the water level. In July, however, after breeding, sedge warblers quite suddenly switch habitat and outdo reed warblers in their devotion to wall-to-wall reeds. The whole population evidently decamps to large beds that provide particularly rich quantities of an insect known as the plum-reed aphid. Reed warblers, meanwhile, take a much broader diet that includes many invertebrates besides aphids. The effect of the sedge warbler's binge-eating of a particular insect is that it puts on enough weight, evidently, to propel it on its long flight, seemingly powered by plum-reed aphid rocket fuel. The reed warbler has the broader diet and is not in the same hurry – or doesn't have the same capacity – to dash southwards.

It could well be that many a songbird at least has the means to make enormously long non-stop flights. Research suggests that quite a number of species make a single flight over the Sahara Desert, for example – apparently at least some swallows do this and as a neat departure from the norm, they fly at night as well as by day. Another piece of evidence comes from species that make Atlantic crossings, many of them by mistake. A few of these lost strays turn up every year in Britain, usually in autumn; many undoubtedly use ships for stopovers, but some probably undertake the whole flight in one go. It has been demonstrated recently that many songbirds from the north-east United States follow the shortest, great circle route from New England to the West Indies and South America, breaking out southwards and traversing the Atlantic for 3,000km (1,860 miles).

It wouldn't take too much of a westerly gale to bring them to our shores.

As mentioned, a handful of songbird migrants get blown across every year. I have a suggestion for you: go and see one if you possibly can. There are certain wildlife marvels that everybody should see: starling roosts, massive seabird colonies, and so on. Another is to go to seek a truly lost soul, just in from the Americas, a songbird a long way from Nashville.

I have had the privilege of such an encounter, when somebody found a black-and-white warbler at Prawle Point in Devon in October 1987. Despite living in London, the thought of reuniting with a delightful bird that I had seen in New York State a few years earlier was irresistible, and I made the 322-km (200-mile) journey to 'twitch' the warbler in an ancient and clapped-out old Mini. I took two friends with me, and it turned out that one was travelling in an unreliable car and had an equally unreliable bladder. His need to stop at every service station between Fleet and Exeter was nearly the undoing of our enterprise.

We arrived at Prawle Point, exhausted but full of anticipation, and we were delighted to hear, upon walking down a gully, that a moment earlier the bird had been in full view of its band of assembled admirers. If you have ever travelled to see a rare bird, you will recognise that the first emotion upon arrival is either relief that the bird is still there and alive, or despair and foreboding that it hasn't been seen. Contented that our chances of 'connecting' with the black-and-white warbler were enhanced by a positive report, we joined the crowd of 50 or so, every one of them with binoculars trained ahead. Several people nodded and pointed. 'It was up by the dead branch a few minutes ago,' advised a bearded man, helpfully.

We trained our binoculars, but the dead branch no longer hosted anything but a few gnats. The surrounding foliage, reddening to a chestnut hue, was still. After 10 minutes, I began to feel frustrated. The people in the crowd that had been admiring the bird began to drift away, off for a celebratory beer. Soon only six of us were left, rueing our missed chance like autograph hunters watching the star's limo drive away down a rainy street. The search area was a small wood, dark with sycamores and hawthorns, and there were plenty of places for a bird to hide for a long time. The black-and-white warbler had already earned itself a reputation for disappearing. After half an hour it was clear that its reputation had been well earned.

I couldn't help myself grumbling that one pee stop fewer and we would have seen it straight away, and could have avoided the task of climbing muddy paths and working our way through the maze of ivy-clad stems on what could be a fruitless mission.

The friend's bladder cost us five hours of searching. However, just as the shadows lengthened on a torturous afternoon, we looked up into the sycamore canopy one more time to see that a small bird had materialised. I immediately recognised it as the black-and-white warbler, simply by the way it was creeping along the horizontal branches, in nuthatch style, a characteristic habit. Given a better view, we could soon admire its smart black stripes on the whitish sides to its belly, and the white wing-bars. The black-and-white warbler has been aptly described as a 'living mint humbug'. Now we had this sweet treat nicely wrapped up. After a couple of minutes, the warbler disappeared; it was never seen again.

But those few moments of seeing the foreigner – how can I describe them? They were triumphant and inspiring. There was a magnificent incongruity of seeing an American bird wild in England. Whatever the bird's journey had been like, it had a remarkable backstory. And of course, the agony of just missing it and the drudgery of the long search, had rendered the eventual success particularly gratifying.

Actually, there is an analogy. A few years later, when my daughter was obsessed with rollercoasters, we went to Alton Towers theme park in Staffordshire on a steaming hot day. We waited under the sun for a particular ride in one of those interminable, good-natured queues that all parents will recognise. The ride broke down, we left the queue, the ride was repaired and we re-joined it. After two hours, with my daughter sunburnt and dehydrated enough to excite the interest of social services, we finally found ourselves strapped into the rocket-propelled ride, with the lights about to change.

The whole ride lasted about 10 seconds. And every millionth of a second was worth it. The two-hour queue was worth it. We were giddy with exhilaration. Seeing the black-and-white warbler was just like that.

CHAPTER NINE

Finding the Way

One of the greatest questions about bird movements – and indeed, within the whole field of all bird study – is the puzzle of how these wild animals find their way. Year on year, birds travel between places with extraordinary accuracy, arriving at regular times and at exactly the same point on the earth's surface. This wondrous mystery has turned out to be something of a jigsaw puzzle, with dozens of different pieces contributing to the whole answer of how they do it. It might go without saying, but most readers are probably aware that many pieces remain to be filled in and that we do not yet have anything close to a complete answer. In this chapter I intend to give a brief summary of this absorbing subject.

The first word should perhaps belong to Professor Ian Newton, who says in his book *The Migration Ecology of Birds*:

'To migrate effectively, birds need a sense of where they are, or need to be, a sense of direction, an ability to navigate from one place to another, and a sense of time, both season and diurnal… In short, they need the equivalents of a map, compass, calendar and clock, together with a good memory, all packed into a brain that in some birds is no bigger than a pea.'

Before touching on the longer journeys that a bird undertakes, it is easy to forget that birds need to know where they are on a very modest scale, too. You and I go to bed in a house, but birds frequently change roost sites, and could wake up in a different place several nights in succession. They need to know where they are just to go foraging, drinking, loafing, and whatever else they do. Birds need to recognise the different parts of their territories and their borders. In short, birds all need a good spatial memory. If they leave their territory for any reason – a predator attack, the need to utilise a rich food source, or to roost – they must remember how to get back. Furthermore, as Professor Newton mentioned, they also need a good sense of time so that they can function effectively.

What, though, of migratory movements, the ones that take a bird well away from a breeding area, say, and propel it to a winter area, perhaps thousands of kilometres away? The first point to make is actually a deeply profound one. Each individual bird has its journey mapped out within its brain, so most of the trip – or at least, the first one it undertakes as a youngster – is purely instinctive. A swallow 'knows' when to set off, when to go south, when to go south-east, and so on. It also instinctively knows when to stop, which is just as well for a swallow that otherwise would carry on beyond the southern coast of Africa to oblivion.

We know this has to be instinctive, just by simple observation. The case of the cuckoo tells us all we need to know. Young cuckoos are brought up by a range of birds, each with different migratory strategies – dunnocks don't migrate; meadow pipits sometimes migrate, but not far; reed warblers fly to West Africa. Cuckoos, however, fly to a different winter place to all three of these species. It is certain, therefore, that they cannot and do not follow their hosts on migration. A young cuckoo gets no guidance from an adult, either, because by the time it fledges, its genetic parents are already deep into Africa. It cannot gain by following another young cuckoo, because every other young cuckoo is in the same boat. These youngsters don't have advice or maps, yet make it to the right wintering areas. They must know the route purely by instinct.

Although the cuckoo is an extreme example, instinctive direction finding seems to be the norm. Many juvenile birds set off on migration after their parents have already left, so once again, unless they can latch onto a tardy adult, they have no experienced bird to follow.

You might counter this assertion by saying, 'Don't birds usually travel in flocks on migration, so they can help each other?' Well perhaps, but if no bird in a flock knew the way, then it would be a case of the blind leading the blind. Besides, we shouldn't be surprised that birds flock together during migration; if each individual knows the route instinctively, then they will inevitably meet up with each other on the way. When we travel from London to, say, Madrid, most of us will go through hubs such as airports, motorway service stations and ports en route. It doesn't mean that we are helping other travellers or have any kind of stake in their journey, we just keep crossing paths. Birds cross paths too, and if a number are crossing the English Channel at the same time, they will flock together for extra safety and comfort. However, in terms of the totality of its journey, every bird is on its own.

There is now plenty of research that indicates that migrants are capable of making their way instinctively and alone. My personal favourite concerns a study on captive garden warblers, which were kept in cages in Germany while their wild colleagues were on migration to Africa. As I already mentioned on page 155, each night the detained birds showed their migration restlessness (*Zugunruhe*) by fluttering around the cage. By measuring the direction in which the caged birds instinctively fluttered, Gwinner and Wiltschko discovered that throughout September the captives tended to want to fly on a south-westerly bearing. By October, though, their preference had changed to south and south-south-east, which exactly corresponds to what a wild bird would do if migrating in Africa. The caged birds were shadowing the movements of wild colleagues, vicariously migrating.

The final proof, if it were needed, that a migration route is built in came from cross-breeding experiments. Breeding blackcaps from southern Germany fly south-west in the autumn, while blackcaps from eastern Austria fly south-east, to avoid the Alps. When migration researcher Peter Berthold and his colleagues interbred captured blackcaps from each population, he found that the general direction of the hybrids was largely to the south – intermediate between the preferences of the parents. This means that, not only is a migration route instinctive, but it is also inherited. When the same experiment was carried out on southern German blackcaps interbred with non-migratory blackcaps from the Cape Verde islands, off West Africa, the hybrids showed migratory behaviour, proving that it had been bred into the offspring of non-migratory birds.

Hopefully, you are now convinced as a reader that, incredible though it may seem, a migrant bird has its migration mapped in its head. That, though, cannot work without external information. A bird 'knows' in which direction to go, but in order to go in that direction, it needs a reference point. Most readers probably already know what birds use as reference points, but the detail of how they do it still bears repetition because quite honestly it is so wondrous.

Let's assume for a moment, then, that a bird wishes to fly south-west. By far the most obvious clue to this is the one that we would use if we didn't have a map or a compass – the sun. So you might say: well, this is dead easy – I could migrate if I was able to use the sun. That's true, although it would confine your migration to the daytime.

It has been shown on a number of occasions that birds use the sun to orient themselves. A classic day migrant is the starling, and it was as long ago as 1951 that Gustav Kramer placed starlings in registration cages (for a description of a 'registration cage', see page 154) and was able to alter the direction in which birds with migration restlessness flew by setting up a series of mirrors to make it appear that the sun was in a different direction. In fact, he made starlings with an inbuilt preference to fly south-west orient north-east instead. Not only did this show that birds use the sun as a type of compass, it also showed that it overrode any other orientation mechanism for this species.

So the birds use the sun to fly south-west or whatever direction. While you can hardly miss the sun, in good weather at least, you will recall that it does have the annoying habit of appearing to move across the sky, from east to west. That obviously makes flying in a particular direction for any length of time much more difficult: a bird would constantly have to re-calibrate its direction relative to the moving object. And of course, it can only do this if it has an accurate idea of the time. If it wanted to move on a bearing of south-west, then obviously the sun would be in a different place in the late afternoon than it was in the morning.

And here we have a startling conclusion to make. Birds must have an accurate internal clock. And this internal clock must adjust for the season. And if a bird is a transcontinental migrant in no particular hurry, such as a reed warbler, the clock will also need to adjust for latitude as the bird creeps south and then north again. And, in March and October, a resident bird needs to recalibrate between GMT and BST by one hour (no, only joking about that last bit, but you wouldn't put it past them).

Clever experiments have duly proven the body clock in racing pigeons. The birds were placed in cages under artificial light and gradually their artificial day was shifted. When released back into the true daylight, their homing abilities were impaired, and they went off course in a fashion that was directly attributable to their incorrect body clock.

Before you begin to think that pigeons and other birds are super-beings that really ought to take over the planet, you might be surprised to hear that their body clock has an inbuilt 'flaw'. Whenever pigeons calibrate time, they don't do so using the height of the sun above the horizon, which we take into account when establishing what time it is – high in the sky towards noon and close to the horizon at dawn or dusk. Instead pigeons, and presumably all other birds, use the azimuth, which is the compass bearing of the sun if you draw an imaginary line down from the sun to the horizon. They cannot distinguish between noon and sunset.

Speaking of sunset, you might imagine that it would be a particularly useful clue in orientation and that is certainly true. Not just because the sun sinks slowly onto the western horizon, but because this is also a time when many birds – the night migrants – begin their journeys, with most setting off 30 to 45 minutes after sunset. In fact, if you consider that some birds will only fly for a couple of hours a night in the early stages of migration, then the point of sunset will still be obvious for some time after they have left. But it turns out that sunset is even more useful than that, because of the effect it has on light. The atmosphere is a weak polarising filter, in that the molecules in the atmosphere restrict the 'flow' of the electromagnetic waves of light in certain directions. Without going into too much detail, this polarisation of light causes a pattern in the sky. At sunset this is particularly clear at 90 degrees to the sunset, and it thus points to the north and can help in orientation during sunset and after the sun has gone down. At least seven species of birds have been shown experimentally to be able to detect patterns of polarised light, which they probably perceive in the ultraviolet spectrum. As yet, though, nobody knows how they use it.

Once night has truly fallen, of course, the stars come out. As long as humans have walked the earth we have had a grand view of the firmament. It is only in our modern world, dogged by light and other pollution and a way of living that often means being shut off from the sky, that we in the developed world are almost immune to the wonders of the night-time light show – imagine what it must have looked like when there was no pollution a few thousand years ago. We are one of only a few generations of humans that have not had an intimate conception of the rotation of the stars. So we shouldn't be surprised at all that birds use a star compass – the pattern of the night sky is no less obvious a celestial cue than our nearest star, the sun.

The classic experiments that first proved birds orient themselves with the night sky were performed in the 1950s in Germany, a time and place of inexhaustible and ingenious migration research. Franz and Eleanor Sauer took a number of birds, including blackcaps and garden warblers, all in a state of migration restlessness, and placed them under the projection of a planetarium. By manipulating the position of the stars on this false sky they were able to alter the direction in which the warblers were compelled to fly, proving that they were able to recognise the stars and their apparent movement.

At first, it was thought that, just as birds might inherit knowledge of their migration route, such birds are born with a map of the stars in their head, so to speak. As it turns out, the way they use the rotation of the stars is far more wondrous – they learn it in their youth.

This was shown in some ingenious experiments carried out at Cornell University in the USA by Stephen Emlen (not a German, despite the name) in the 1960s. His test subjects were not British birds, but indigo buntings (they have been recorded in the wild in the UK, so I can include them here!) Emlen hand-reared several groups of indigo buntings. The first group were kept in a windowless room with diffused lighting. When they were placed in registration cages inside a planetarium showing the northern hemisphere night sky in the autumn, they showed appropriate migration restlessness but did not orient in any particular direction. The second hand-reared group were reared in a windowless room without seeing the sun, but every other day for two months of their youth they were placed in a planetarium showing the northern hemisphere night sky, with a simulated rotation around Polaris, the north star, as you would see in nature. In the autumn, these birds showed migration restlessness and flew away from the north star, exactly as they would have done in the wild.

The really clever part of the experiment was the treatment of the third group. These were reared in exactly the same way as the second group, except that in their youth, they were shown a falsified picture of the night sky rotation. Rather than rotating around Polaris, the firmament was modified to rotate around the star Betelgeuse, in the constellation of Orion, which is in the south-west part of the sky, about as far from Polaris as you can get. When the restless youngsters, eager to migrate, were placed in registration cages under the correct northern hemisphere sky, rotating around Polaris, their inclination was to fly away from Betelgeuse and almost towards Polaris, in the opposite direction. This proved beyond doubt that indigo buntings, at least, learn the rotation of the stars in the period after they hatch and before they migrate.

Subsequent studies have shown that northern hemisphere migrants fundamentally learn the rotation about Polaris, but also learn the location and pattern of a number of other nearby (in relative terms!) constellations, including the Big Dipper, the Little Dipper, Draco, Cepheus and Cassiopeia. Incidentally, there happens to be a British bird called the dipper, but it is one of the most sedentary birds we have, with little or no migration in all our populations. So it seems that the feathered dipper might not learn where the Big Dipper in the sky is!

The marvel of all this, to me at least, is the fact that, during the summer, young birds must surely spend some time awake at their roosts, watching the stars rotating in the night sky. What a marvellous thought.

The fact that birds use the stars to fly in the right direction on their migratory flight spawns an interesting question. Do birds fly at night in order to use the stars, or did the habit of flying at night evolve first, with an ability to use the stars second? After all, if you think about it, songbirds don't as a rule, fly around at night, so what makes the night attractive? We will never be able to answer this chicken-and-egg question, of course, but one thing is clear: the night holds some real advantages for birds flying long distance.

The first advantage is that at night-time the air is cooler than in the daytime, reducing the threat of dehydration and heat stress while aloft. Of course, if this is critically important, every bird would migrate only at night, unless it needed to soar on thermals, as a bird of prey or stork might. This cannot, therefore, be the only answer, although it might allow some extra comfort. Another argument against the importance of air temperature is that, as we've discussed, most birds set off about half an hour after sunset. If they really wished to benefit from cooler conditions, you might expect them to set off later.

Another hypothesis, which like the previous one cannot readily be proved or disproved, is that atmospheric conditions are kinder at night. There is less turbulence, and this particularly suits the rather slow, not very powerful birds that make up the nocturnal migratory community. If you combine this with the cool, pleasant conditions, that would be a powerful incentive to move at night.

One more suggestion is equally impossible to prove. If birds migrate by night, then obviously that conveniently leaves the daytime for them to attend to their other needs, such as feeding, preening and resting. Most of the long-distance night migrants are insectivorous, and because insects are easier to catch by day, it stands to reason that the birds could feed up by day and migrate at night. The only snag with this hypothesis is that it really only fits birds that are in a hurry. As we've seen, it isn't at all unusual for a bird to fly for a short while and then rest for a day or two before moving on. Most migrants, at least in autumn, are migrating in a fairly leisurely fashion, at least at first. So it is unlikely that many of them are compelled to follow a migrate-by-night, rest-by-day, migrate-by-night pattern.

Whatever the reason – and it is probably a typically biologically messy combination of reasons – many birds do move by night. Less than an hour into an August or September night, and sometimes into October, there is a highway above rooftop height that, in the right conditions (not very windy, no rain), will be busy with birds. While we might rave and marvel at huge bird flocks, such as waders on estuaries, there is an equally impressive but invisible phenomenon going on above our heads on still autumn nights. The number of birds suddenly surges after sunset and increases steadily until midnight. It then declines hour by hour, with many birds stopping their migratory stint and landing, still in the dark, in a totally unknown place – although it seems that birds can recognise suitable habitat from the air. Few night migrants are still moving at dawn; most will have landed and will be starting their first meal as soon as there is light in the sky.

Alas, we cannot witness this. Perhaps if we took a hot-air balloon and rose no higher than 1,500m (4,920ft), we might see dozens of birds passing in the night air, calling quietly. What an experience this would be. If we had a powerful torch, we could illuminate them, and maybe attract them in, and we could see them face to face, wild birds on the move. We might be able to witness streams of individuals and small flocks, of many species, probably pulsing in waves, and perhaps for hours at a time. We could watch the great migratory procession of birds on the move, meeting up with familiar names that we have never seen before. It would blow our minds.

It is possible, just, to get an idea of what is happening. There are two ways to do it. You can set up a telescope against a full moon and occasionally you will be able to see small shapes passing. Another, easier way is to wait for a very still night and then simply put your ears to the sky. Just sometimes, you can hear something. Your best chance is to hear the high-pitched 'see' and chuckling 'shack, shack' of redwings and fieldfares respectively, mainly in October. However, here in Dorset, some pioneering birders have taken recently to placing a sound-recording device on top of buildings and leaving it on all night to record bird traffic. It is extraordinary what they have picked up, although it is mainly ducks and waders. Coots and moorhens have been a feature. But they also have picked up songbirds such as pied flycatchers and blackbirds and wagtails on the move. Who knows what might pitch up next? The point is that, if you know your bird calls, the night sky is limitless.

We all know that a bird's ability to orient itself is a marvel, but no discovery magnifies our sense of awe and wonder more than knowing that birds can use magnetic fields. This is something spine-tingling, completely alien to us. Imagine the bird up in the sky, sensing nothing more than the truly minute, weak magnetic forces of the earth. Our planet is like a giant compass, with field lines of magnetism running from south to north. At the magnetic poles, the field lines incline downwards at 90 degrees, while at the Equator they run horizontally for many hundreds of kilometres. In addition, there are variations in magnetic field intensity all over the world, in a pattern that is unique to each location. This is stuff humans can measure, but cannot perceive. The imagination goes wild.

I have already mentioned (in chapter 7) that the magnetic sense of a bird has been confirmed by experiments on robins. So far, it has been further tested for about twenty species of birds. In every case so far, the birds have what you might call a magnetic-inclination compass – in other words, they perceive the lines of inclination into the earth, and apparently not their direction. So they cannot tell that the lines run from south to north, but they can tell how far they are towards the pole or how close to the Equator. If the lines are almost flat they are near the Equator, while the steeper they incline towards the earth, the further north the bird is. This obviously works for birds both in the northern and southern hemisphere. Near to the Equator, any migrants would have to use other cues.

The Holy Grail (or at least, one Holy Grail) of bird-migration studies is to find out exactly what physiological quirks grant birds their migratory sense. At the moment there are two suggestions, which might well work together. Both account for the fact that, although the earth's magnetic field is entirely predictable and reliable, detecting its weak signal would require extraordinary sensitivity. One possible source is the known presence of iron oxide (Fe_3O_4), or magnetite, in parts of a bird's brain. This chemical will become magnetised when exposed to a magnetic field, and it changes shape slightly, which could set off a signal to the brain. The other possibility lies in a chemical reaction within a protein called cryptochrome. When exposed to short-wavelength (blue) light, cryptochrome becomes activated, forming two free radicals (molecules with an unpaired electron). The magnetic field so-formed affects how long the cryptochrome remains activated. It is thought that cryptochrome might be present in retinal cells and affect their light perception, effectively allowing the birds to 'see' the magnetic field. Some birds are known to lose their magnetic sense if their retina is damaged.

The idea that birds could see the magnetic field feels like science fiction. They might as well be super-birds and use kryptonite.

This recent discovery that birds might be able to see the earth's magnetic field puts an interesting complexion on some classic experiments performed on robins in the 1960s (see pages 154–5). The birds were put in registration cages to measure their preferred direction of flight, and when they were studying whether magnetic field had an influence on their orientation, the scientists put large electrical coils in with the birds. Quite what they might have seen with their eyes as they contemplated these strange magnetic contraptions is open to question.

The second Holy Grail of bird migration, and the mystery that should earn a Nobel Prize for whoever convincingly solves it, is how birds actually navigate from a completely unknown location to where they wish to go. This, you will appreciate, is a much more mysterious ability than pure orientation, which is the ability to fly in a preferred direction from a definite starting point. Yet somehow, as we know, if you put a pigeon on a plane and fly it to somewhere completely novel, hundreds or thousands of kilometres away, it can, on occasion at least, find its way back home. Other birds have been shown to be able to do this, too. After many hours shut off from their usual cues, they are launched, bleary-eyed, into the air of a location that they have never visited before, never seen before and not planned to visit. Yet somehow they manage to work out where they are relative to their home loft.

Surely the only answer is that these birds have a kind of map in their brain. We can only speculate about what it is, and how they use it. There are wonders inside the head of a pigeon – and we tend to think they are stupid!

One more question. What migratory abilities do birds possess when they don't actually migrate very far, if at all? Returning to a few paragraphs back, can a dipper appreciate the Big Dipper? Do house sparrows, which hardly move anywhere, have a magnetic sense like obligate migrants? If a blackcap can change its migration route, can it become a migrant? Do non-migrants have the battery of remarkable senses used by their migrant colleagues, which are often closely related genetically? From the tropical rainforests of Amazonia to the dry forests of Australia, there are birds that never stray more than a few hundred metres from where they were born – indeed, the ranges of some species of antbirds (*Thamnophilidae*) are delineated by rivers, because these shade-loving birds are unwilling to cross a couple of hundred metres of water. Could these same birds find their way around using the sun, stars and magnetism?

The chances are that every bird species has the latent ability. Consider the case of racing pigeons, mentioned above. These animals defy science by being able to return home after being displaced inside a lorry or aeroplane and sometimes released thousands of kilometres from their home loft. Yet their ancestral form, the rock dove, which used to breed on the wild cliffs of Scotland's Western Isles and is now mainly found in unpolluted form in the deserts of the Sahara, is entirely resident. The motley crew of pigeons on your neighbour's roof may themselves stay put in your cul-de-sac for their entire lives, never once venturing beyond the parish boundary. If they can home, they can migrate.

So, while migration is a marvel, it is also probably universal. To birds, indeed, it is routine drudgery. To us, it is another world.

CHAPTER TEN

Recreating the *Pastoral Symphony*

In 1808 the composer Ludwig van Beethoven completed work on his sixth symphony, and one of his most celebrated pieces of music. The vivid *Pastoral Symphony* is a mellow paean to the countryside. At the end of the second movement, the *Scene by the Brook*, is a very clear depiction of three bird sounds – a nightingale, a quail and a cuckoo. We know for sure about the composer's intentions, because the symphony had unusually clear programme notes for the time, and the three names *Nachtigall*, *Wachtel* and *Kuckuck* are unambiguously labelled. The nightingale song, represented by a flute, is distinctly slowed down and somewhat imaginative, while the quail (oboe) and cuckoo (two clarinets) are, to a birdwatcher's ear, in perfect rhythm. The last note of the quail's 'wet-me-lips' overlaps the first note of the cuckoo's song, which is exactly the sort of juxtaposition that happens in the outdoors, and leads one to conclude that Beethoven didn't just have joyful feelings about being in the countryside, but was at home there and heard these birds often.

That there should be these three particular species together amplifies this impression of natural history experience and expertise. The species mix is not so unlikely as to be scientifically discordant. Had Beethoven jumbled together, say, a capercaillie from a coniferous forest with a tree sparrow from a rural village, or combined a summer bird like a swallow with a winter visitor such as a brambling, we could accept it purely as artistic licence. But he didn't. Hearing a nightingale, a quail and a cuckoo together is possible. Tellingly, the sequence at the end of the movement is only a brief snatch and the composer doesn't develop it or embellish it beyond what is almost empirical mimicry. It is therefore natural to convince ourselves that, one day, Beethoven was indeed walking beside a brook and heard the overlapping quail and cuckoo and stored it away in his mind. He was well known for taking long constitutionals in the countryside around Vienna, where he was working at the time, and frequently took notes. What I'm leading to is this: the *Scene by the Brook* could easily have been a real experience. True, the species chosen are highly symbolic of love (nightingale), of God's provision (quail) and the coming of spring (cuckoo). But bear with me; they can just as easily have been living and breathing at the time.

Interestingly, while it is possible to hear those three species at the same time, it is not a routine juxtaposition and would have been noeworthy 200 years ago, too. That's because of the birds' slightly different habitat requirements. If there was indeed a real *Scene by the Brook*, there would have needed to be a copse of trees and bushes beside the water for the nightingale, and the brook should have been adjacent to a meadow or field for the quail (cuckoos have broader habitat requirements). The point is, however, that it needs a specific set of circumstances to hear the three sounds at the same time. If an experienced birder had heard such a chorus, it would have immediately raised interest. You can imagine the experienced country walker Beethoven saying to himself: 'Good heavens, I haven't heard that combination before', and duly noting it down.

If this convinces you that the *Scene by the Brook* was a kind of musical reportage, and Beethoven really did hear the nightingale, quail and cuckoo together, then it is surprising how much you can surmise from the event. Not only can you guess the habitat juxtaposition, you can also make a decent stab at choosing the week of the year when this happened, the time of day, and even some of the fortunes of the birds concerned. Once again, when birds open their mouths, they betray a remarkable amount about themselves.

So let's begin with the time of year. All three species are summer visitors. The arrival times of cuckoo into Central Europe peak in late April, and until the end of May the male's song can still be a dominant sound. The anonymous poem tells us that 'their tune changes in June.' This isn't strictly true because their tune can change at any time, but June lies right at the end of their singing season when activity is already tailing off, with very few chances to mate and no territory to defend.

The peak singing periods tell us far more. For example, male nightingales stop singing very quickly after pairing. In Austria, their song output peaks at the end of April, when the females drop in, and declines before mid-May; there is a short peak near the end of the month and then the birds almost fall silent. There can be the odd resurgence, but it would be very unusual to hear one in July, and even then it is probably a youngster trying out singing for the first time.

The fact that quails also first arrive in May should begin to narrow down Beethoven's time zone. A few turn up at the end of April, but the first two weeks of May see the main arrival. It is actually then possible to hear them at any time until August, because of later immigration. However, nightingales and cuckoos have fallen silent well before then. The chances are that Beethoven's trio were all making their sounds in the first week of May.

Another interesting conundrum is the time of day. All three species happen to vocalise more frequently at dawn, dusk and at night than many a bird. Indeed, the quail is usually easiest to hear around midnight, at least in Britain. We would not assume that Beethoven would wander around the night-time countryside near Vienna at midnight in the early 1800s, with few lights and the chance of accident or crime. He was probably walking in the early morning or evening. Once again, the fact that all three species are typically singers at low light authenticates the music.

From this evidence, the *Scene by the Brook* probably occurred near Vienna at dusk in the first week of May. The quail was not yet paired (they stop singing once they have found a mate), but the nightingale probably was. The cuckoo's female partners would probably have parasitised nests of pied wagtails, garden warblers or robins, their favoured hosts in Central Europe in that habitat.

Ever since I became aware of the *Scene by the Brook*, I have mused on the possibly of replicating it in my own experience, to hear all three species calling at once. Yet I have found it astonishingly difficult, especially since I live in Britain, on the edge of the range of nightingale and quail. The timing has never worked, and I have only ever scored two in a day. Cuckoos and nightingales are actually quite easy to hear in chorus but the quail, which here has a penchant for wide agricultural fields, especially of peas, is a different matter entirely. Indeed, hearing a quail in Britain is a major feat, and seeing it is virtually impossible; in 40 years of birding I have never set eyes on a common quail anywhere at all. Quite apart from the fact that it isn't easy to get to chalky fields at night, in the middle of summer when the dark comes so late, you would also have to do so near to a patch of scrub that hosts nightingales, while they are still singing.

Setting aside aspirations to hear all three species, everyone should try to experience hearing a nightingale in the wild. It is one of those wonders that is on many bucket lists, along with jumping out of aeroplanes and visiting the Taj Mahal. Hearing a nightingale is very much cheaper than visiting the Taj Mahal and less potentially dangerous than jumping out of a plane, even with a parachute on. But in England, as mentioned above, the nightingale is localised and you must go during the narrow window when the males are singing. It's actually better to go with somebody who knows the song. In the fresh evening springtime air, sprinkled with sweet scents, it is easy to get confused with song thrushes, robins and goodness knows what else.

I have many times had appointments with English nightingales, usually in the company of groups of people who have come to hear what all the fuss is about. Even before you begin, the fact the spring day is maturing into the evening always gives the trip a special flavour. Few people are familiar with being in the countryside at night time: its darkening shadows, the strange sounds that emanate from somewhere at a distance you cannot judge, the profusion of scents. In the scrub and coppice where nightingales breed you often find ranks of bluebells, which lose their colour as the night closes in. May is often the time when the blackthorn blossom vies for attention with the hawthorn blossom, a battle of two abundant blooms which drape over their parent plant like flumes of froth. The spring day hums with activity, while the night ticks over; the cool stills the ardour, but you can feel it resting.

There is something elemental about daring to be outside in the dark. There was a time, of course, when it was truly dangerous, and even now, in our safe lives, the dark sends a frisson of remembrance. These days it simply adds to the pleasure. We know that if we slip a few steps away from our friends and colleagues we can escape to the cover of darkness where we don't know what lurks. This mildest of temptations is a thrill.

Into this mix of emotions comes the song of birds. Whatever the functional reality of birdsong, as I have already covered in this book, we can put it aside in the midst of an evening spring chorus, and simply revel in the luxuriant sounds. On a spring evening as the light fades, you might expect a reverse of the dawn chorus and, in many ways, that is exactly what happens. The true day-singing birds, such as chiffchaffs and goldfinches, have long fallen silent by the time the light is fading, and you are down to the twilight mainstays, such as robins, blackbirds and thrushes. In one of my favourite nightingale venues, Bookham Common in Surrey (now sadly degraded), there are always a few blackcaps and garden warblers singing at dusk, too. These are fine songsters in their own right; like a nightingale, the garden warbler is a short springtime treat, its singing season over almost before it starts, like a starts. In May the garden warbler emits a delicate, even babble with a slight whiff of mania, while the blackcap utters a flutier offering, always strongest at the end of its phrase. The blackcap sounds like a human walking through the woods whistling; certainly not a musical whistler, more a jobbing whistler, such as a decorator or builder, whistling at the expense of any tune.

The warm-up songsters, therefore, sing loudly in the fading light as if they were the main event, which to their own kind they are. It is a curious thought that a song thrush might be utterly immune to its next door nightingale, but apart from registering each other's sound, the two species have no language in common, so far as we know. Anyhow, the last songsters embellish the evening, particularly as the world as a whole quietens.

There is always a gasp and a smile when the first nightingale starts, typically before the rest of the supporting cast have fallen silent. There is relief all round, the star of the show is on station! It is easy to identify; the nightingale often gives one of its loud, liquid trills early on, and it really is quite unlike anything else. Throughout the songs this evening it will be apparent that nightingales are loud and shouty; they might be top-of-the-range sopranos, but they are noisy sopranos and dramatic sopranos. What sets them apart, to my ear at least, is their dynamic range, which has no equal. One moment their phrase contains delicate notes that you have to strain to hear; once your attention is duly focused, there is a dramatic crescendo, with maximum effort. Who knows what another nightingale makes of the soft and loud bits; to us they are magnificent.

The other aspect that makes nightingale song special is the long trills, which I have mentioned earlier (page 34). These are an important part of a male's repertoire, because the number of sustained trills (fast repetitions of a note) gives listeners a clue as to the male's quality. Once again, although other birds trill, the nightingale gives a more exuberant and confident effort.

Arguably the nightingale's quintessential turn of phrase is the slow, sobbing crescendo to a thrilling, rat-ta-tat flourish. It is dangerous to compare human and bird 'music', but this element of a nightingale's song really could be placed at the end of a piece if this flourish is to be heard to best effect. Although it is fanciful to suggest that this is exactly what Beethoven heard at his famous brook, there is also no reason to assume that it wasn't. As the nightingale was a common bird in Beethoven's home, he would undoubtedly have heard its famous flourish many times and would have remembered the structure and timing of it long after he became deaf.

Night has fallen now on our nightingale trip, and the song perches are in darkness. As we had hoped, all other voices have been quelled by the blackness and the nightingales are performing solo, as a species. Three are singing, close enough to be on each other's vocal toes. It sounds sweet but smells of rivalry.

For a while, we listen in silence, part delight, part deference. What kind of bird is this to produce such a complex song? How did the vocalisation evolve into a biological masterpiece – and from an average brown bird? That's the thing with nightingales. You cannot listen to them without a degree of wonder, or bafflement, a sense of not taking it all in: 'Sounds overflow the listener's brain, so sweet that joy is almost pain' as Shelley put it. And it is not settling, any more than listening to a great human performer is necessarily relaxing. No, you wouldn't put a nightingale on a loop to get you to sleep.

We leave as the moon and the mosquitoes rise, leaving the birds to a contest that looks set to last all night (although it doesn't). I can tell that those people I have brought to hear the nightingale are satisfied, and several are delighted. Nobody feels that the song is overrated, though many say that they prefer various other songs – particularly the blackbird, the less demanding, reassuring song of summer evenings at home. In all the trips I have made to listen for nightingales – made in all weathers, including the cold and the rain – there has never been crushing disappointment. Part of the reason for that is that it is hard not to be beguiled by spring and the spring evening, but the rest is down to one of our most remarkable songsters.

The cuckoo's song could hardly be compared to the nightingale's in any technical way. It is simply two notes repeated in a short sequence, a sound that is standard throughout the cuckoo's vast range, from western Europe east to Japan. Yet to human ears, it is deeply evocative, almost as much as a nightingale. It is similar in some way to the hoot of a tawny owl, which carries a message of danger and menace that has percolated into our culture from our fear of darkness and death. The cuckoo isn't like that, but it is associated with a kind of breakdown of normality; firstly the breakdown of a relationship between male and female (the word cuckolded comes from a cuckoo) and latterly the breakdown of the relationship between young and parent. It carries simple mysteries – where does that sound emanate from? – and deeper ones. There is a questioning air to the cuckoo, and it fills every atmosphere with its strength of volume and far-carrying pitch.

These days, as I have mentioned above, the cuckoo carries another message of breakdown. Its voice has fallen silent over parts of the UK, especially in the south. It has become difficult to hear a cuckoo. My two children grew up on the edge of a Dorset heath, and on a few blessed occasions they woke up with the sound of a distant cuckoo. But how many children in the future will be able to have the same experience?

Beethoven would be astonished today to hear about the decline of all three species in recent decades. There can be no doubt at all that, around Vienna in the early nineteenth century, the song of the cuckoo would have dominated the landscape like nothing else, and certainly outside the experience of most people in Britain today. It has disappeared from our landscape as surely as church bells. The most recent estimates suggest that cuckoos declined in the UK by 49 per cent between 1995 and 2010, and by 63 per cent in England in that time. This is a serious recent trend, a potential catastrophe for the bird and those of us who love to hear the sound. Nobody is sure why, but it could be to do with changing conditions in Africa during our winter, or perhaps a large reduction in the number of moths caterpillars.

The nightingale, meanwhile, is also declining in Britain. Its range has contracted by 43 per cent since 1972, and numbers have fallen by 90 per cent in the last 40 years, of which 52 per cent has occurred since 1995. Again this could be to do with problems on the wintering grounds, as well as a reduction in good habitat here. Meanwhile, the quail hasn't declined here to any known extent, but in Europe as a whole, it has done, massively. The enormous falls recorded in the first years of the nineteenth century (see below) are a thing of the past.

While many of the younger generation have never had the privilege of hearing a cuckoo, probably the majority of people who have heard one have not set eyes on one, which adds to the bird's mystique. There is a technique for seeing them, though, and that is to wait for a calling to end a sequence of 'cuckoo' songs, and then be especially vigilant. Watch out for a hawk-like bird with pointed wings and a long tail; it will be flying below the horizon or just above it, but never high. The wing-beats are fast and quite odd – the wings never look as though they are lifted above the horizontal. After a short while, the bird will land and probably settle lower than you think – perhaps on a side branch of a tree, rather than the top.

I have had some glorious experiences with cuckoos. A few years ago I was sound-recording in Provence, in France, and when a cuckoo began to sing I thought I would take the opportunity to get it on tape (yes, tape!). Just as I turned on the recorder, the cuckoo suddenly seemed to forget how to sing. It managed the 'cuck' alright, but the 'oo' was all over the place. It was all discordant and deflated, a sore-throated failure. It was so funny that I ruined the first part of the recording by laughing.

Another, more recent experience came at a famous birding location, the Belovezhskaya Pushcha in Belarus, a World Heritage Site famous for the primeval state of the trees, which are huge and untouched, left to rot where they fall. The Pushcha is a very wet place and on one morning inside the forest, I was treated to a unique chorus: cuckoo and fire-bellied toad. These animals make sounds of similar pitch, although the toads don't have an obvious double note as the cuckoos do. However, taken together, the forest resounded to magical 'oohs' and 'cucks', which echoed hauntingly and were intermixed with loud and confident chaffinch flourishes. It was a moment to bottle and keep. It was a moment made for Beethoven to recreate with the woodwind.

I haven't personally heard many quails, but I know that these whispering birds also create magical moments. One day, when I was living and working in west London, I heard about a bird that had been calling in a field in Dorset, near the village of Sixpenny Handley, two hours' drive away. On a whim, I decided to drive there after work. The sun was just setting over a vast field on gently rolling hills, and the atmosphere was perfectly still. The sun turned into an orange ball and the smells of summer wafted all around me. Within a few minutes, there was a gentle 'wet-me-lips, wet-me-lips' from a few metres away. I listened for quite some time, almost until dark, but it didn't call again. Yet the encounter took me into the deep countryside, into a world I only flitted in and out of occasionally, and the call of the quail was intoxicating. Many years later I have settled in the county of Dorset, not so far from the quail, but a world away from west London.

I have never heard all of Beethoven's birds singing at once, but I have at least managed the next best thing: hearing them all on the same day. This was back in 1996, and although it didn't seem significant at the time, the location was the eastern lowlands of Austria, not far at all from Vienna. I could have walked in the composer's footsteps, although Lake Neusiedl, where I heard the birds, would have been a long walk indeed. I can remember hearing cuckoos every day, and seeing them often. Nightingales, too, were singing from all the scrubby areas. Once or twice I came upon a quail, once in a large weed patch within a path in the middle of the day. The weed patch was thick, but only a couple of metres wide. It was overgrown with poppies and cornflowers and types of yellow cress, and I tried to flush the quail by walking through it. I didn't succeed the first time and felt guilty about doing it again, so the bird remained an unseen enigma. But Central Europe caught my imagination: the architecture, the imperious musical heritage, the thought of the Danube flowing east to the mysterious Black Sea, formerly behind the Iron Curtain. The birdsong was so loud that it felt as though it was smoking from the bushes and trees, the leaves and blossom were at their best, the sun was out each day. And the hostel where I stayed was staffed by two sisters in their thirties, both blonde-locked and glamorous to the hilt, delicious, bewitching and Germanic, as completely alluring as they were out of my league. Never mind; I felt supremely alive.

Being a decent naturalist, by inference at least, Beethoven would have been astonished to learn about the lifestyles of the three birds he reproduced in the *Pastoral Symphony*. His was a time when people tended to think of nature as much purer than it really is – indeed, it is only recently that many of the gorier and less palatable aspects of the natural world have been discovered. It is worth remembering that nobody had yet worked out the breeding cycle of the cuckoo in Beethoven's time, let alone the remarkable secret life of the quail and the complicated society of nightingales. Nobody had yet worked out why birds actually sang. You could imagine the fun and games that the great composer might have had if he had known about them – the turbulent and evocative pieces he might have been inspired to write. The *Pastoral Symphony* would still have been sweet and relatively light, but it might have been more convoluted.

Having said that, the cuckoo's unusual nature has been known since the time of Pliny the Elder, at least. In the first century he recorded that smaller birds were eaten alive by cuckoos, although we now realise that this was a graphic description of the foster parent stuffing food into the juvenile cuckoo's enormous mouth – the cuckoo doesn't eat birds, only insects. It wasn't until Beethoven's time that the first-known accurate account was published of the nestling cuckoo ejecting its nest mates, either eggs or chicks, using the small of its back to heave them over the side. This was in 1788, and the observer was the multi-talented Edward Jenner, who went on to develop the world's first vaccine, for smallpox. Many of the details remained elusive until a classic study by Edgar Chance in Worcestershire between 1918 and 1925; he was the first person to confirm that the female laid its eggs directly into the host nest. These days the cuckoo's story is so familiar that it is easy to forget just how extraordinary it is.

I don't wish to labour over a life history that is well known to many, but there are a few points about the cuckoo, and its subcontracting of the effort of parenting to a completely unrelated foster-species, that might be worth mentioning, because they illustrate further just how sophisticated the lives of birds can be.

The first point has only been discovered since DNA analysis became a standard tool in science. And that is that each individual female cuckoo becomes a parasite only to the species that raised her in the nest. To put it another way, a female cuckoo brought up by reed warblers will be a professional layer of eggs only in the nests of reed warblers, throughout her breeding life. Her eggs will usually mimic that species, and so will those of her female descendants. The tendency to lay eggs in the nest of a particular species is passed on along the female genetic line, while male cuckoos are thought to copulate with a female of any persuasion if the opportunity arises. There are, therefore, genetically different 'meadow pipit cuckoos', 'reed warbler cuckoos', 'dunnock cuckoos' and 'robin cuckoos' in Britain, although they are physically indistinguishable. They aren't considered to be species as yet, nor subspecies; instead the different female lines are known as 'gentes'. Away from Britain, there are many other types of host. Some of these are only casual with no genetic exclusivity, probably because a female cuckoo has run out of host nests and thus tries to lay in a different nest to normal (a cuckoo can lay up to 20 eggs in a season).

You will doubtless be well aware of the cuckoo's lifestyle, but have you ever wondered how cuckoos actually find nests? And at the right time? After all, there is no point in a female cuckoo laying an egg in a nest in which the chicks have already hatched, or laying an egg before the host has begun its clutch. On both occasions the egg would come to grief, being outcompeted by the host's own young in the first instance and being removed, or causing nest abandonment, in the second. In fact, cuckoos must time their parasitism very acutely. If they get it even slightly wrong, their breeding season will be a flop.

To prevent this happening, cuckoos are avid birdwatchers. They wheedle their way into the territory of the host and silently watch the comings and goings of pairs from a concealed position. If a member of the host species discovers the hiding place during its working day, it will call loudly and harass the larger bird relentlessly, often attracting other species of birds with the commotion. However, as often as not the cuckoo sits unmolested and mentally takes notes of what is happening. It must need to know where the nests are and what stage each pair is at. It must always lay its egg, one per nest, when the host species has started, but not completed the clutch. That way, when the owners return to their parasitised nest, they will find the same number of eggs as when they left – the only difference is that their own egg has been eaten or destroyed, and the cuckoo egg is well placed to do its worst.

Another crucial aspect of timing is to enter the host nest when nobody is at home. For this reason, cuckoos are very unusual in that they lay their eggs in the afternoon. The morning is a busy time for a songbird. The female of the host species will lay an egg at dawn and needs to feed up. The pair will probably copulate. There is much to engage their attention. In the afternoon there is less to do, maybe some preening and more feeding. The females won't be sitting yet. On the whole, small birds don't begin incubating until they have finished their clutch or there is just one egg to go. That way the eggs will hatch more or less all at the same time, allowing them all to develop at the same rate – indeed, the cuckoo takes advantage of this because its egg is already partially developed in the oviduct, and thus is already ahead of any host eggs. The point is, though, that there will be times in the afternoon when the nest is left alone. The cuckoo, however, will only be able to take the opportunity if it has been concentrating keenly.

Another remarkable aspect of cuckoo biology is that the female manages to find the host nests at all. With reed warblers, the cuckoo's main host in Britain, that is reasonably easy. The warblers prefer to nest close to others of their kind, in clumps if not colonies, and they always nest within reeds above water, a very restricted, distinctive and structured habitat. The nests are woven between the reed stems, using the latter as struts; they are well protected from predators that don't wish to get their feet wet or climb reed stems, but they aren't well protected from cuckoos. You and I, quite frankly, could parasitise reed warbler nests if we were small, laid eggs and had wings.

The nests of the other hosts, on the other hand, are much more difficult to find, mainly because they are much more dispersed – the cuckoo must work to uncover them. Robins nest in cavities, sometimes out of reach of a cuckoo, so they are the least common host in Britain. Dunnocks build cup-shaped nests in bushes and shrubs. Meadow pipits build on the ground, often in the long grass in open areas including pasture, heath and moor.

It seems that the pipits sometimes give cuckoos a helping hand in finding out their own secrets. On occasion, the cuckoo comes out of hiding and confronts a pair of meadow pipits in the open. This provokes an entirely natural, reflex response: the pair of meadow pipits angrily mob the cuckoos and give a series of anxiety calls. The trouble seems to be that the meadow pipits cannot help themselves. Any appearance of a cuckoo will be stressful, but the nearer the cuckoo is to their nest, the angrier and more agitated is the pipits' response. It is like a game of 'hotter, colder', whereby a human being leads another to a hidden item by saying 'hotter' if you are moving towards it and 'colder' if you are going in the opposite direction. In bird society, however, this is not a game. The cuckoo can monitor the pipits' response and find the nest, led by the hosts inadvertently.

Why don't the pipits cotton on to this and deliberately confuse the cuckoo by being quiet? In the arms race that wages every year between cuckoos and their hosts, perhaps one day they will.

There is one addendum I would like to add to the cuckoo story. Their method of reproduction is called brood parasitism. They never breed any other way but off-letting all parental duties to a completely unrelated host – many other species of cuckoos in other countries actually build their own nests and raise their own brood. What cuckoos do is very unusual, although not unique; there are a decent handful of other bird families that also do it, most of which have just one host species. However, there is another kind of brood parasitism that is much commoner and could even be happening in your garden. It happens when a female of one species lays eggs in the nest of its own species – in fact, a neighbour. This is called intraspecific (within species) brood parasitism, and it is carried out quite frequently by, for example, starlings.

Intraspecific brood parasitism is a very cost-effective way of increasing your reproductive potential. As a pair, you defend a territory, build a nest and lay eggs as usual, but in addition, the female of the pair steals into the neighbours' nest and 'dumps' one of her own eggs in it. In the case of starlings it probably works because these birds nest in cavities and might have trouble recognising their own eggs or knowing how many there were (odd, knowing how vital this information is) in the gloom. The practice is not as abusive as the cuckoo's parasitism because the marauding neighbour doesn't usually destroy a host egg. Nevertheless, it is cheeky, and clearly, it sometimes works.

While nineteenth-century naturalists (and composers) had perhaps just an inkling that the cuckoo's life was unusual, they would have known more firmly about the peculiarities of the quail. As mentioned above, these small birds are extraordinarily secretive, at least when breeding. During migration, however, they do some pretty strange things, things that have been recorded for thousands of years. In the Bible, they are mentioned as turning up in large numbers, seemingly out of nowhere, and literally covering the ground (there are Roman accounts of similar happenings). For centuries they have been caught for food in the Mediterranean region, and there are records from Italy of many thousands arriving overnight. In 1918 nearly 2 million quail were exported from Egypt. The quail was a bird people knew well enough to eat. It was as good to feast upon as the pheasants and partridges to which it was related, but would fly strongly and fall miraculously into the hunter's fold.

But the quail's migration is even stranger than the ancients could have imagined, at least in the context of what we know about other birds, and the details have only fairly recently been discovered. Essentially the quail's migratory strategy is similar to what we see in butterflies. It seems that most years quails are not present in Europe in winter but are resident in North Africa (just as are, for example, painted lady butterflies). Here they begin to breed in March or April, while others from further away migrate into Europe as relatively conventional summer visitors. Then, having bred once, all or part of the North African population suddenly moves north into Europe – this includes the young that have only just recently hatched. The second invasion takes the quail into Central Europe, or even into northern Europe, where some may reach Scandinavia. Here the adults breed again and the youngsters breed for the first time, even though they could be just 12 to 15 weeks old.

Unpredictable, small, elusive and relatively silent, quails are among the most difficult of all wild birds to study. Only now have we learnt what goes on in the hayfields, so to speak. It seems that the distinctive song attracts birds migrating overhead, both males and females, and therefore quails form shared 'mating centres' that could be in the middle of a large field, but nowhere near any other quail. Within these communal breeding areas, the males hold no territory, but the song is sufficient to attract a female, and the pair bond is monogamous.

And so to nightingales – what strange behaviour might they have hidden away in their locker? Compared to the previous two species, they are models of decorum, with males only occasionally taking two mates. However, in the spirit of these times, their relationships are notably short, only lasting a single season. But the nightingale's life is not one of Shakespearean intrigue.

However, Beethoven would have approved of this: it has been discovered that nightingales have two different songs, not one. One type they sing during the dawn chorus, evening chorus and off and on throughout the day; the other is sung only during the night. To be honest, it is quite difficult to distinguish between the songs without analysing them carefully, but the daytime song is not as rich as the night-time song. It is less variable and has shorter phrases and briefer pauses between them. And the two different songs, it turns out, have quite different functions. The day song is the usual territorial skirmish, while the night song is for mate attraction. You can imagine any composer's relish at the thought of two types of song, and how he or she might be able to interweave the different phrases.

Fascinatingly, scientists have radio-tracked female nightingales to prove the differing motivations of the two songs. What they find is that, as soon as the females have arrived on the breeding grounds, invariably later than the males, they spend a number of nights trawling around the neighbourhood, listening to the potential talent. Soon birds pair up, and the lucky males then cease singing much except during the dawn and evening chorus. Once the females have paired, they obviously stop their night-time reconnoitring. Bachelor males, however, sing all through the night in the hope of attracting a non-breeding female or possibly a late arrival. The females tend not to be very active at dawn, and so the dawn chorus is devoted to territorial posturing, male to male. Quite why female nightingales look or listen for a mate by night is not known.

There is one more question to muse upon concerning the possibility of replicating the *Scene by the Brook*, and this subject will concern us for the rest of this chapter. The question is: if you found a spot with a nightingale, a quail and a cuckoo all singing at the same time, even if it was near Vienna, would they actually all sound the same as they did when Beethoven heard them in the late eighteenth and early nineteenth centuries. In the 200 years since the Sixth Symphony was written, has birdsong changed? We know that the instruments used for music do sound slightly different these days (and we also know that the *Pastoral* debuted badly in 1808, with an ill-prepared orchestra and a bored audience) – but do the birds? The wider question is this: does birdsong change with time? And you can then ask: if it does change, how profoundly does it change? What would a nightingale, a quail and a cuckoo have sounded like 200 years ago, 500 years ago, 1,000 years ago? How has birdsong evolved? How might the song of a nightingale have evolved? And if birdsong does change, why does it change? What are the selective pressures that might cause it to alter over the course of time?

It isn't possible to answer any of these questions without delving into how a bird acquires its song. Is the song in the genes and, if so, does it come out rough or polished? And what is the mechanism for a young bird beginning to sing for the first time?

The answers to some of these questions will probably be obvious from previous sections. After all, chapter 2 describes how singing male birds are judged on their songs – whether it be repertoire or song rate – and how their song affects their fortunes in the breeding season. We learnt that different males produce individually distinct songs. How can this happen?

In fact, the songs of birds have complicated origins. Some are simply inherited and come as a package with the bird – the cuckoo is an example of this, and so is the quail. The more complicated songs, though, must have a way of developing into a repertoire that is unique to the individual. This could, in theory, come from a set of pre-determined parameters passed genetically – in other words, the song could be inherited as a unique characteristic expressed from a unique recombination of genes. However, for a number of reasons, this is unlikely – not least because if a song varies too much it might not sound like the right species. In practice, most bird species do at least inherit a sort of 'aural template' that prevents this. To allow for a degree of conformity and a degree of inventiveness, however, most birds actually go through a period of learning their songs. They do so by listening to the songs of adult birds in the near vicinity, and then there is a period afterwards, dependent on the species, when they can embellish a song into something that has its own personal signature.

The first experiments to prove that a bird song can be learned were performed on chaffinches. William Thorpe hand-reared young birds in the 1950s in Cambridge, exposing them to different acoustic environments. One set of youngsters was reared in auditory isolation, the fledglings hearing no birdsong at all between the spring they were reared and the following spring. When it came to the time when they should begin singing, they produced a pale shadow of the usual trill and flourish, lasting the correct time (1.5–3.5 seconds) and in the right frequency range, but lacking clarity and form. Some individuals left out the terminal flourish altogether.

In contrast, birds reared in isolation but exposed to the tape-recorded song of a chaffinch managed to sing perfectly well the following season. This was the case even when they had only heard it months earlier. They remembered what they had learnt.

Another set of youngsters were reared together, but also without adults for tutoring. They did better than those in isolation, but their rudimentary songs were still undeveloped and sounded similar to each other. In common with teenagers everywhere, they listened to their peers, copied their peers and the result was a mess. Other chaffinches were played other bird songs, and even human instruments, but none of these helped unless they were strikingly chaffinch-like. It was abundantly clear that, unless a chaffinch hears another chaffinch singing 'properly' in its youth, it will not be able to produce the correct song when it grows up.

Since Thorpe's time, many other birds have been tested. Some won't even learn from a good recording of their voice, but only from a live tutor. Others, including reed warblers, seem to be able to make up for lost time by re-learning later in life. A few seem to keep learning throughout their lives. However, the general rule seems to be that birds are born with a species-specific template in their minds. They listen to tutors in early development, during a sensitive phase which varies in length between fifteen days and many months. For a while they are completely silent apart from quiet moments when they mumble their song under the breath and then, the following spring, they test their attempted songs against adult birds in the same vicinity.

You might think that witnessing this song-learning behaviour is out of reach of the ordinary listener, but that is far from the case. Most birdwatchers and garden enthusiasts have undoubtedly heard birds practising and trying to perfect their songs. There are two times of year when you are particularly likely to hear this going on – during the middle of winter and early in the breeding season, in late January and February. And the chaffinch is, indeed, a good one to listen for. They start singing in earnest rather later than other birds, several weeks later than great tits, for instance. You might hear the first attempts in late January, but chaffinches don't fill the air with their short, cheerful phrases until close to the end of February.

The amusing thing is that you really wouldn't think that the chaffinch's simple phrase would require much learning or practice. After all, it is essentially a rattle that accelerates slightly towards the end until it ends with a bang. It isn't quite Tchaikovsky's *1812 Overture* ending with its fireworks and drums, but it is the loudest part of the song. The whole phrase lasts up to three and a half seconds.

Yet if you listen to chaffinches singing in early February it is hilarious. They seem to make a complete mess of their simple task, getting everything wrong you can imagine. They often leave off the terminal flourish; they sing too slowly, and become hesitant. They start badly, give up after a few notes and try again. Listen enough and you begin rooting for them, and if they finally produce something vaguely similar to the correct song, you just want to cheer.

This incidentally brings to mind the strange case of the hopeless collared dove. Collared doves, as many suburban gardeners know only too well, have a somewhat dreary song which is three notes in extent, and approximates to the rhythm of a football fan droning 'United', repeated a few times in succession. Apparently, although I cannot claim to have heard this myself, some singers are incapable of completing the phrases, singing just two notes instead of three. You would hardly expect that singing three notes would be beyond them, but it is evidently so, or perhaps they are simply lazy. Anyhow, the reaction of listening females to this is entirely predictable. The inadequate singers spend a frustrated season wondering what they are doing wrong.

Another opportunity to hear birds practising occurs at sporadic times in the autumn and winter. In the wild, to hear so-called 'subsong' is a rare treat because it is the birth of something greater, like the first draft of a great piece of writing. I have heard robins indulging in subsong in autumn and blackbirds in the middle of winter. In the case of the robin, I had no idea at first what I was hearing. In place of the very regular pattern of a robin – a phrase, long pause, a quite different phrase, long pause, yet another different phrase, and so on – there was nothing but a very low and quiet babbling. For any birders reading this, it sounded rather like a very quiet whitethroat song. The bird was vocalising from low down in an evergreen, and it took a good 10 minutes to spot the bird and establish its identity. Part of the problem was that, when the sound came, the robin didn't open its bill. Only when I managed to creep up to about a metre away did it become clear that it did indeed emanate from a robin. The bird seemed unusually relaxed. There was no sense of challenge, no proclamation. It seemed to be sung under the breath.

When I have heard blackbirds subsinging, it is more obvious what they are. As in the case of the robin, the main stream of sounds had a babbling quality, but this time there were quite obvious blackbird phrases intermixed. In common with the robin, the bird had its bill closed as it sung, perhaps to avoid calling the attention of other blackbirds. This was on a warm December evening, well in advance of the usual blackbird song period, which begins in February like the chaffinch. On this occasion, in a quiet back street in a provincial town, the performance sounded exactly like a muted rehearsal.

For a brief period of his childhood, my son Samuel went through a similar process. With a happily loquacious elder sister sharing the house, he would often be quiet in the daytime. However, after bedtime, alone in his room, for a while he would chatter to himself, laugh and even sing the theme tune to *Bob the Builder* being his favourite – before falling asleep. Admittedly he did have his mouth open when doing this, but in other respects, his practising of language was not so different from the robin and the blackbird. Several recent scientific papers have drawn close parallels between the way birds and humans learn their vocalisations.

The fact that individual songbirds learn from their fathers, usually as their main tutor, as well as from their father's rivals and neighbours, has several interesting repercussions. One of these is that songs can have a local flavour. Imagine that a chaffinch with a 'twee' note at the end of its song, acquired by dint of its own inventiveness, arrives on a small island off Sweden one spring. All goes well and it pairs up with a female and produces three sons. These three sons will learn its song type with the 'twee', as well as the father's other song types (he may have six) and some song types of his father's neighbours. The next season the surviving sons will sing the 'twee', as might other young males who also picked up the father's unusual repertoire. Within a few generations, the island could be populated by chaffinches that sing the 'twee', while on the mainland there are no such irregularities. What you are left with is a culture in one place where chaffinches add the 'twee', different from what occurs elsewhere.

The scenario described above occurs in nature, and it has one immediate useful practical application for the birds. Whenever male birds settle into an area to breed, either after being in winter flocks or arriving from their spring migration, they sing in order to defend their territories (chapter 1). At first, their songs induce aggressive responses from their neighbours, but it doesn't take long before they get used to each other. One way of doing this is to sing similar songs. In song duels, they may use exactly the same phrases: one might sing the avian equivalent of 'You're an idiot' to its rival, eliciting the response 'You're an idiot'. They trade blows and, over time, there is an extremely localised language. Even within birds with highly complex songs, with hundreds of sound types, this so-called 'matched counter-singing' occurs. It is of particular value when the territory is visited by a stranger. The stranger will be easy to recognise because it sings a different song, making it easier to detect, easier to find and, hopefully, easier to evict.

These song environments are equivalent to the human existence of dialects, and they may be found over very narrow areas or very wide ones. The redwing, which is a common winter visitor but a very rare breeder in Britain, shows a very unusual pattern of dialects. In southern Norway, researchers found the average area of a dialect was only 41.5km^2 (16.02 square miles). The boundaries were very sharply defined year on year, and within a boundary most males sang only one song type. If they sang two, such birds held territory close to the border with another dialect. Much is remarkable about this, not least that the birds would evacuate the area in winter to go to another part of Europe, often different from year to year, yet return to this small area defined by sound. Most birds contain their dialects in much broader, larger areas.

If birds have dialects, as people do, the possibility arises that we humans might be able to recognise them. In some cases, with a suitably good ear and aural memory, you can do it, especially if the bird only sings a simple song. One very good example concerns one of the songs we have been looking at already, the chaffinch. In Britain, chaffinches sing their accelerating phrase with a terminal flourish; in parts of the continent they do the same, but add a single syllable at the end, a sort of 'tic'. Being at the end of the phrase, you cannot miss it. You have picked up a chaffinch dialect.

Real experts on the song of one particular species can recognise exactly where it comes from. Imagine that. If you were similarly expert, you could pick out a Welsh robin, a Geordie blackbird and a south London song thrush. That would be quite a party trick. Of course, at a dinner table with friends or neighbours there is nothing to stop you pretending you know.

Birds don't only have dialects; on a much wider scale, they might also need to tailor a song to their habitat. Great tits singing in thick woodland usually have lower-pitched songs than their counterparts in thinner woodland or gardens, simply because the sound transmits better through leaves using the relatively longer wavelengths of lower pitch. Their songs also contain more pure tones, and overall have a simpler structure. This distinction holds good for the community of birds at large: birds of more open habitats have higher-pitched songs.

There is another important biological by-product from birds learning or part-learning their songs and that is that, over time, the songs themselves could change. If, for example, a bird gradually adapted to a new habitat, it might be advantageous to alter its song to suit the conditions. If this happened at a pace where the female population adapted their preferences, you could imagine natural selection favouring a new song. If it happened little by little, it would never change enough to sound like a different species, so the whole population would still recognise it. But if you took two points in time and measured the change, the song could eventually be unrecognisable.

We don't need to make any assumptions that birdsong changes with time because we know for sure that it does.

A number of experiments have shown this, a good example being an American study on a bird called the savannah sparrow. In this case, researchers compared the birds singing on Kent Island, New Brunswick, in 1980 to those singing in 2011. The birds had added more clicking notes into the middle of their rather dry phrase and changed the ending. Apparently, only the first part of the song identifies them as savannah sparrows, and this didn't change much, but cultural changes affected the middle and the end.

A more subtle piece of evidence comes from studies of chaffinches on the Canary Islands and the Azores. Chaffinches have only invaded these offshore islands within the last half million years, but already their songs have changed enough to be quite difficult to recognise. Many of the elements are the same, but the order in which they are sung, the syntax of the sentence if you like, breaks down.

You can probably see where this chapter is going. *The Pastoral Symphony* was first performed more than 200 years ago. Since that time, many human generations have come and gone, but many more nightingale generations have come and gone. Musical instruments have changed, and some sound different today to how they sounded in Beethoven's time. Human speech has also changed. If you try to read poetry from 200 years ago, there will be many archaic words. Shakespeare from 400 years ago requires study, Chaucer from 600 years ago requires translation, yet both are English. If you and I were transported back in time to the fourteenth century, we wouldn't understand much of what was spoken to us. Human language evolves, so it is reasonable to assume that birdsong does too.

From this we can arrive at a startling conclusion. If we had been in Beethoven's shoes at the *Scene by the Brook*, would we have recognised the nightingale? The cuckoo is maintained perfectly as the note of E in the symphony, as it sounds in nature, and the quail is one of those birds that doesn't learn its song, but inherits it. The quail and the cuckoo would probably have sounded exactly as they sound now, but for the nightingale, we can only speculate.

There is a somewhat disturbing conclusion to this story. Changes in bird songs haven't only been uncovered in wild situations; they have also been recorded in reaction to our human world. Birds as diverse as blackbirds and great tits are now known to sing at higher pitches in cities than they do in more rural locations. This enables them to be heard better above the roar of traffic and other human noise, and it is also apparently related to the architecture of the cityscape – higher frequencies adapt better to the peculiar distortions off buildings and other man-made structures. According to one study, this need to change to a higher pitch could affect the great tit's breeding success, because females tune in particularly to low-pitched songs when they are at their most fertile. And who knows what other abnormalities there are in the singing life of birds? Never before has there been so much noise with which to compete. It is a far cry from the world of 200 years ago.

The truth is, of course, that we can never quite return to the *Scene at the Brook*. If Beethoven's brook was near Vienna, it might well have been heard to a backdrop of church bells and the hubbub of human life – indeed, there are passages in the music where the rhythmic digging and hoeing of the peasants finds its way into the score. But the amount of peace and quiet in the world was at a level that we would never recognise today. There would have been no radios, no cars, no trains yet and no passenger planes, the latter the major cause of irritation to bird-sound recordists, along with wind noise. As we will see later, the sheer abundance of wildlife, especially in the still-rich farmland of Central Europe, was at a level none of us see any more, especially in Britain. The soundscape that Beethoven experienced in his early years, before deafness blighted his life, would have been at many levels richer than we are used to. It wasn't a perfect world, of course; it was blighted by wars, famine and disease. But it would be hard to imagine now the degree to which the world was a quieter place.

So you cannot recreate Beethoven's *Scene by the Brook*, however much you might try to seek out the three species singing together. The world has changed too much. And perhaps the nightingale has changed, too.

CHAPTER ELEVEN

A Thousand Cuts

In the garden of the suburban house where I lived as a boy in south-west London, a pair of carrion crows cast their shadow. They used to nest each year at the top of a giant oak that towered over the neighbourhood. The tree seemed to my youthful eyes to be 100 or more feet tall when it was probably only 40 or 50, but it was a suitable platform from which our high-rise crows could lord it over the rest of the birds. It seemed to me that the body of the garden avifauna was petrified of the crows, as they would scatter from all the bird feeders theatrically with a flap of the jet-black wings. Once I saw the crows attacking another bird on our garden lawn, the smaller bird's wings fluttering desperately beneath the sleek assassin's body. The struggling stilled and the crow flew off with a puff of feathers; it was a despotic act from a bird at the peak of its powers.

Every morning there would be a brief but very loud blaring of caws as the crows arose. The pattern was a volley of yelled sergeant-major caws, which I always assumed came from the male, and these could have blown dust off the surface of a table. Then there would be a response with a little less edge, which I assumed was the feminine line. The call and response were prompt, and everybody knew that these two were completely in league, the fiercest element in this suburban oasis.

Year after year this couple built their stick nest in the branches of the giant oak. It was the highest nest in the neighbourhood, well above the colony of jackdaws in the rooftop chimneys. The smaller birds would chatter and fuss whenever the crows went anywhere near them. The crow's nest was always completed early, well before the leaves came out, hence it was easy to observe the comings and goings at the penthouse suite. The nests withstood all the wind and rain of the early season and most years the pair successfully raised a family of chicks.

However, overnight on one fateful day, everything changed for our crows. It was October 1987 and, during the Great Storm mid month, the huge oak tree was blown down. The view over our garden was permanently changed, and in a small way so was my birding life, since visits to my family house were a little less birdy. However, the effect on the crows was far more devastating. Until that point, the pair of crows in Richmond Park Road never ceased to hold their command, on top and impregnable. When they lost the canopy, they lost everything.

The next year they attempted to breed of course – at least, I assume that they were the same pair. No tree in the neighbourhood was anything near as tall as the oak, which was once a standard in a pre-war orchard. There were some very old fruit trees, and in the following years crows even tried to nest in the canopy of an ancient and rotting apple tree nearby. The nest, though, was never as large or robust as the old one high above ground, and the birds never raised young from it. In fact, I am pretty sure that the pair never again raised young in our neighbourhood. The jackdaws were probably much less stressed, but for the crows their legs had been taken out from under them.

You probably don't have any sympathy for my crows. I'm not sure whether I did at the time. The garden lost its edge, though. It lost a little biodiversity. I left the area shortly afterwards, and when I returned for a visit years later the jackdaws had gone too, along with the swifts and house martins. My birding home had been absorbed into 'anywhere else, London'. Nearby Richmond Park, enclosed by Charles I in 1637, provided a home for delightful small birds called redstarts in my childhood, and I have photographs of tree sparrows from the side of Pen Ponds, in the centre of the park. Both these species have also gone in the last thirty years.

Why tell the story of the stricken crows? It is told because it is a story of personal disaster. True, the disaster was entirely natural, and countless similar stories remain untold every single breeding season, for all kinds of birds. However, the real point is that, once the crows lost their territory, they lost their capacity to reproduce. Once stricken, the alternatives didn't work. If their new territory didn't provide what they needed, then it is doubtful that any other crows would have moved in once it became vacant. Crow society is quite polarised and, once the pair gave up on their territory, they probably joined a non-breeding flock and saw out their days without ever reproducing again. This is the reality for birds that lose territories to circumstance rather than to rivals. They go quietly and aren't replaced. Without the big oak, the breeding crow population reduced by two.

It so happens that many years later I moved to Ferndown, in Dorset, a relatively countrified area. In 2015 the trees that used to house our nearest rookery were cut down to make way for a housing development. The rooks were part of the fabric of the town. Any time in the spring and summer that

I drove south towards Poole, I would check the rooks and see what stage of the breeding cycle they had reached. In January, they would fly over the road with sticks for refurbishment; in March they would fly in with full throat pouches for their youngsters; in June there would be rook nurseries on the nearby fields, where well-grown youngsters would unashamedly beg for food. Walk past the rookery and you would be 'serenaded' with the wide variety of these birds' noises, not all harsh and deep but many musical and high pitched, trebles within the caws. They were timeless sounds and sights, part of the countryside's fabric. Now they have been moved.

Who knows, perhaps the rookery has successfully relocated. Without colour-marking the birds, who can tell? A small part of Ferndown has been changed, though; the treetop serenade will be replaced by the sound of mowers and laughing children and cars starting up. None of these are bad, of course, but it would be better to have rooks *and* bricks, not just the latter. The kids in the new housing estate won't have the perky corvids to create their childhood atmosphere.

There are other parts of town that have changed. Some nearby allotments between a wood and a golf course have also been turned into a relatively swanky, but totally anonymous housing estate. The same thing happened to a local wood before we moved down there. A field where roe deer now frolic, close to suburbia, is earmarked for another big development.

Just along the road from our house was a vacant plot of land. Nobody seemed to know why a house hadn't been built there, but every year for the 16 years we lived in Ferndown the plot became more and more overgrown. A significantly large buddleia took root, and it became a butterfly haven. It was the quintessential neglected corner, content not to be loved or noticed. Now there is a superstore outlet in the place where everyday invertebrates crawled over everyday earth and munched the leaves of everyday plants. Nobody cares that it has gone because the creatures that lived there are entirely unremarkable; the buddleia isn't even a native plant. In the scheme of things, the loss of neglected land put to good use is hardly a tragedy.

However, Ferndown is just a small town, and what has happened in the last few years is hardly unusual; the tide of shrinking wild and neglected land is ubiquitous. The conversion of $10m^2$ ($108ft^2$) of untidiness into a large shop is hardly objectionable; the selling off of allotments for housing is understandable; the loss of the rookery is regrettable; the destruction of a local wood is irredeemable. Each represents a different scale of change, but add them together and the loss is unconscionable. And that is the problem. In our land wildlife is suffering death by a thousand planning decisions.

Make no mistake, we are talking deaths, not relocations. Ecologically illiterate planners and politicians may speak of wildlife 'going elsewhere' when land is concreted over, but this is a myth. You can relocate individual plants and you can offset destruction, but in most cases, these things don't happen, particularly in small, overlooked untidy corners. An animal deprived of its habitat dies.

Take the example of a wood full of nightingales, one of the *causes cèlèbres* of recent times in Lodge Hill, former Ministry of Defence land in North Kent. If you cut down the woodland for a housing development you won't directly kill your nightingales. If they find their breeding areas have disappeared when they return from wintering in Africa, they will, of course, be displaced and relocate. The problem is, there isn't some Shangri-La where a suitable nearby woodland has plenty of space simply crying out for the secretive birds. The refugees may or may not find a new wood to their liking; if not they simply won't breed. If they do they will have competition. The nightingales who bred there or hatched at this 'new' wood last year have probably already arrived and set up territory, and they won't relish the extra bodies fighting over the same space as last year. The interlopers will be at a disadvantage. Many will miss a year of breeding. Some of the incumbent birds will lose their territories. Over the course of time, the 'new' wood will return to holding the same number of nightingales as it did previously, although some of the new birds may survive. However, whatever happens there will be a net loss of nightingales to the overall British population.

Many people don't understand this, because when people are displaced the expectation is that they will be absorbed into a population somewhere. In the harsh reality of a bird's existence, there are no safety nets.

Does this matter? Do individual cases matter? For most people they don't matter – that Marks & Spencer will improve the town, and the property developers' firm will benefit from the new build and it will be a pleasant place to live. In that sense, it doesn't matter. As people, we understand the need for commodities. A large majority of the population could probably live with a concreted-over countryside. They wouldn't miss rooks or butterflies.

The problem is that big things start small, and the individual wildlife losses mount up. In Ferndown, it isn't so much the housing estate on the allotment that is the problem, it is the fact that people get used to new housing estates in places like this, and more will be built. The more used to it we get, the more space will be used up. If you are worried about healthy eating, you will recognise that, if you allow yourself one more biscuit, you will do the same tomorrow, until you expect a biscuit every time. Each biscuit on its own is harmless, but over the course of time, eat enough biscuits and your health could suffer. Indeed, your health may pass a tipping point that you never notice, and the consequences will be dire.

I am saying that development breeds development. New developments are harmless enough in isolation. But their totality adds up to massive, creeping changes in our world, and in what's left for wildlife. It seems now that the totality of creeping changes in our landscape might have reached a tipping point, or might even have passed a tipping point, where we are about to suffer severe losses. These will be isolated at first – birds, butterflies, crayfish – but if the losses carry on, they will begin to hurt us.

When I was young, much of the attention and money in conservation was lavished on high-profile cases, such as ospreys, red kites and avocets, birds that were famous for their rarity and imminent disappearance. In that simpler world, many of the battles were won: all three species still flourish in the country. Nowadays, partly because of improved surveying techniques, conservationists' fears are now focused on the fortunes of common species. While red kites soar, birds such as mistle thrushes are in serious decline. And the falls in songbird populations, in particular, are causing despair.

Take the case of the skylark. When the first detailed 10-km^2 (3.86-square-mile) survey of British breeding birds was undertaken in the mid-1960s, the skylark turned out to be the most widely distributed bird, found over 98 per cent of the whole country, from the arable fields of England to Scottish moorland. Between the years of 1970 and 2010, the population is estimated to have fallen by 58 per cent, and it has been lost from about 10 per cent of its range.

In 2015 there was an online poll conducted to choose Britain's National Bird. The robin was the obvious winner and other serially popular species, such as kingfisher and mute swan, received a healthy number of votes. The skylark, however, did not make the top ten. This is the skylark, a bird that has inspired poetry, prose and music. It is a bird that pours out its free-flowing song from high in the sky, showering the land with exuberance, spraying vocal gold like a living firework. This is a bird that used to form the atmosphere of a hundred summer picnics. You could walk over downland and farmland for a day, and when you put your tired head upon your pillow that evening the sound would still ring in your ears. The skylark was abundant but iconic, a small brown bird that everybody adored.

Why, then, did it not come near the top of the poll? The fear is that the skylark is no longer well known enough to enter into people's minds and affections. Simply not enough people have heard it, been beguiled by it and found their heart lifted. The whole land hasn't fallen silent, but it has quietened. Each small corner that has lost its lark equals one fewer opportunity for a listener to hear one. The retreat of the lark hasn't been one big catastrophe, but the accumulation of small setbacks, each one barely noticed or not noticed at all. The upshot is that the skylark has retreated from our consciousness.

The problem with skylarks isn't just about skylarks, of course. Skylarks don't live in isolation, and they can be seen as a symptom of a greater malaise. If it was only skylarks, we could reluctantly, but satisfactorily attribute their decline to a quirk in their particular biology. Even then, their decline – unless it was caused by disease or some kind of mutation – would inevitably be related to something in their ecology – the insects they ate in the summer or the seeds they foraged in the winter. In other words, a skylark's decline is never just a skylark's. Something else in the environment is always involved.

The trouble is that the skylark's decline isn't isolated. True, there are birds that are doing fine in Britain, such as great spotted woodpeckers and blackcaps. But it seems there is a critical mass of birds in trouble. Up until recently the farmland species have been monopolising the bad news – yellowhammers, corn buntings, tree sparrows and grey partridges have all declined steeply. However, in the last decade or so, it seems that the skittles are falling everywhere: house sparrows in towns, down 69 per cent between 1977 and 2010, kestrels in open country, down 44 per cent since

1970 and, perhaps most worrying of all, a suite of woodland birds are in trouble.

I have been birdwatching long enough to have noticed some of these declines first hand. I saw one of the last cirl buntings breeding in West Sussex in the 1970s and have seen some of the last willow tits still to breed in Surrey and north Hampshire in the 1990s. But none have affected me in the same way as the crash in the population of lesser spotted woodpeckers. It has become a personal sadness.

The lesser spotted woodpecker is a delightful bird, in every way a mini version of the great spotted woodpecker – indeed put the latter in the washing machine at too high a temperature and you might end up with the former. The lesser spotted is well named whichever way you look at it: it is the lesser of the two species of spotted woodpeckers in this country, but it is also spotted less often. Until fairly recently, the lesser spotted woodpecker would be a bird that you came across fitfully if you were regularly outside in woodland with small trees, or along rivers or canals with plenty of alders. In the 1990s and early 2000s I used to encounter it often, at least once a month. I have watched, enthralled, as two female lesser spots fought over the affections of a male, and I have watched as one member of a pair fed a single chick up in the tall canopy of a New Forest wood. I have heard them call and drum, in their petite way. I have even observed this little bird in my Dorset garden, although just as an occasional waif.

Now I never see them. Indeed, two years have passed since my last encounter. I have probably lost the ability to spot them, but it would be extraordinary indeed if my experience was simply a matter of overlooking a bird I was good at finding. And sure enough, the lesser spotted woodpecker is indeed disappearing. It has gone from many of its former haunts – I used to see it frequently in Bushy Park in south London. I used to see it in my garden in Ferndown, but not any more. The population collapsed from about 1980, down 63 per cent by 1998, and it has been falling precipitously since. There may be fewer than a thousand pairs left in Britain – and nobody is sure why.

These days there is a desperation about losing some of our commoner birds. It comes from a feeling of increasing impotence, that the world we know – in the form of government and people at large – is not listening and does not care. However, the problems for songbirds are very much greater than the problems that we personally give them as British citizens. Migrant songbirds face a whole raft of other human-induced problems.

In the spring of 2016, I took a walk in the New Forest with a small group of birders. It was May, and the spring was hyperventilating with breeding activity, set in a backdrop of impossibly vivid green from millions of oak trees. Most of the Forest's special birds were easy to see. There were redstarts, robin-like sprites with orange tails that shiver all the time as if the birds are trying to shake something from the tip. Tree pipits sang their chaffinch-like songs from the tops of trees, too lazy under the grey skies to launch into a song-flight. In every holly tree, there seemed to be a firecrest, the Forest's new buzz-bird and a wondrous success story of increase, and siskins flew between the tops of the spruces. Goodness me, we even saw a pair of hawfinches that looked as though they had a nest in a tall alder tree. Hawfinches are stupendously shy and difficult to see, so we thought it couldn't get better – until a still rarer honey-buzzard flapped regally over, on its way to a secret nest-site.

The forest was rocking, as we say these days. With a little of the afternoon left there was just one bird left that I was expecting to see (not the lesser spotted woodpecker) and I knew exactly where to go to find perfect habitat. We followed a large, well-maintained cycle track down past fields with horses, down the edge of a heath and finally slipped down into something of a New Forest wonderland.

This National Park is a little short of good bluebell woods, but here, next to a small brook, bluebells covered the ground all around us. These weren't the deeply intense, overcrowded bluebell woods you normally go to see; they were more spread out and wispy, as if the woodland floor was covered in green-blue mist. Unusually, these bluebells were underneath beech trees, the latter newly dressed to the nines with whorls of crispy lime-green leaves. With their negligible understorey and airy canopy, and each with distinct strata of leaves from the browse line to the top, these beeches were gift-wrapped for a delightful New Forest special, the wood warbler.

I listened out for the wood warbler's absolutely distinctive song, a shivering, accelerating trill that, with a little imagination, can be likened to the sound of a spinning coin coming to rest. In between the shivers, the bird utters equally distinctive 'pew-pew' notes, also in a slightly accelerating series. The French name is a euphonious *pouillot siffleur*, the latter word meaning 'whistler'. Wood warblers are unique among British forest birds in performing a songflight inside the canopy – not a very impressive one, admittedly, but it adds up to make the wood warbler the quintessence of effervescence, if that isn't a touch fanciful. These small birds are also colourful, deeply buttery yellow on the face and throat, brilliant moss green on the wings and tail, and creamy white on the breast. They are, quite literally, bright and breezy.

We listened for 40 minutes or more, and not a single wood warbler sang. The season, the habitat and the history all pointed to them being there, but they were not.

Speaking to New Forest experts, I have heard that our population has crashed – down by more than 50 per cent in a handful of years. And it isn't just here, but everywhere. In Britain as a whole the wood warbler has reduced in range by 37 per cent between 1988–91 and 2007–11. Within this range numbers dropped by 65 per cent between 1995 and 2010.

Even in continental Europe, the species has been in decline at least since 1980.

Nobody is entirely certain about what is happening to wood warblers, but the most likely explanation is that its wintering grounds in tropical West Africa are being degraded by human activity.

Here, then, is a bird that we can help a little, but cannot help very much. It is a small bird whose biggest problems are far, far away. I cannot see how we can begin to get wood warblers back in numbers, throbbing in the spring leaves. My deep regret at this only intensifies my sense of impotence.

And how about birds that are facing ruin through climate change, potentially one of the greatest problems humanity has ever faced?

This should compel us to act when, within reasonable parameters, we can have an influence. There is another bird whose story should be told, perhaps for no other reason than that it shames our inaction. The case of the turtle dove is one that discredits a foreign country. It shames civilised society in that country. And it shames us for our lack of proper anger. Here in Britain the turtle dove has declined by 90 per cent between 1997 and 2010, and in Europe as a whole it has declined by 73 per cent between 1980 and 2010. Yet on the island of Malta, a member of the European Union, the government allowed 5,000 to be shot every year, a direct violation of the EU Bird Directive. Although there is now (as of May 2016) a moratorium on turtle dove shooting, how many of the shooting community on that island care?

It seems that not even the might of the European Union can stop a small country allowing the killing of a species widely known to be in freefall. So what chance will birds have in the future?

At the moment, birds are losing ground. It is death by a thousand development plans, changes in farming practice, concreting over of lawns, and a perfect storm of greater threats still. Somehow, we must reverse the trend. The only solution, I suspect, is by thousands of people caring enough.

As for the birds themselves, isn't it time we helped? As we have seen in this book, they have enough to contend with in the natural course of their lives as it is.

Acknowledgements

Writing a book can be a solitary and selfish activity, so thanks are owed to Carolyn, my wife, and my children Emmie and Sam, who had to cope at times with an absorbed, distracted husband and father. They did so admirably, with love and grace.

Many thanks to my commissioning editor at Bloomsbury, Lisa Thomas, who got the idea accepted and signed off. Thanks to Anna MacDiarmid and Julie Bailey for taking the project through to its conclusion.

Finally, many thanks to Marianne Taylor for her astute editing and for her delightful line drawings that adorn the chapter headings.

Bibliography

General Bibliography

Balmer, D. *et al* (eds). *Bird Atlas 2007–11: The Breeding and Wintering Birds of Britain and Ireland*. Thetford, Norfolk: BTO Books. 2013.

Berthold, P. *Bird Migration: A General Survey*. Oxford: OUP. 1993.

Catchpole, C. & Slater, P. *Bird Song: Biological Themes and Variations*. Cambridge: CUP. 1995.

Cocker, M. & Mabey, R. *Birds Britannica*. London: Chatto & Windus. 2005.

Cornell Laboratory of Ornithology. *Handbook of Bird Biology, Second Edition*. Princeton University Press. 2004.

Cramp, S. *et al* (eds). *Handbook of the Birds of Europe, the Middle East and North Africa, Volume 4 – Terns to Woodpeckers*. Oxford: OUP. 1985.

Cramp, S. *et al* (eds). *Handbook of the Birds of Europe, the Middle East and North Africa, Volume 5 – Tyrant Flycatchers to Thrushes*. Oxford: OUP. 1988.

Cramp, S. & Brooks, D. J. (eds). *Handbook of the Birds of Europe, the Middle East and North Africa, Volume 6 – Warblers*. Oxford: OUP. 1992.

Cramp, S. & Perrins, C. M. (eds). *Handbook of the Birds of Europe, the Middle East and North Africa, Volume 7 – Flycatchers to Shrikes*. Oxford: OUP. 1993.

Cramp, S. & Perrins, C. M. (eds). *Handbook of the Birds of Europe, the Middle East and North Africa, Volume 9 – Buntings and New World Warblers*. Oxford: OUP. 1994.

Del Hoyo, J., Elliott, A. & Christie, D. (eds). *Handbook of the Birds of the World, Volume 10 – Cuckoo-shrikes to Thrushes*. Barcelona: Lynx Edicions. 2005.

Michl, G. *A Birders' Guide to the Behaviour of European and North American Birds*. Budapest: Gavia Science. 2003.

Perrins, C. *British Tits*. London: Collins. 1979.

Scott, G. *Essential Ornithology*. Oxford: OUP. 2010.

Steen, M. *The Lives and Times of the Great Composers*. London: Icon Books. 2010.

Wernham, C. *et al*. *The Migration Atlas: Movements of the Birds of Britain and Ireland*. London: T & A. D. Poyser. 2002.

Key References

Chapter 1 – Awakenings

K. S. Berg, R. T. Brumfield & V. Apanius. Phylogenetic and ecological determinants of the neotropical dawn chorus. *Proceedings of the Royal Society*, 2006.

T. J. Brown & P. Handford. Why birds sing at dawn: the role of consistent song transmission. *Ibis*, 2003.

R. S. Hartley & R. A. J. Suthers. Airflow and pressure during canary song: direct evidence for mini-breaths. *Comparative Physiology*, 1989.

A. Poesel, H. P. Kunc, K. Foerster & A. Johnsen. Early birds are sexy: male age, dawn song and extra-pair paternity in blue tits, *Cyanistes* (formerly *Parus*) *caeruleus*. *Animal Behaviour, 2006*

Rothenburg, D. *Why Birds Sing*. London: Allen Lane. 2005.

K. A. Schmidt & K. L. Belinsky. Voices in the dark: predation risk by owls influences dusk singing in a diurnal passerine. *Behavioral Ecology and Sociobiology*, 2013.

R. J. Thomas *et al.* Eye size in birds and the timing of song at dawn. *Proceedings of the Royal Society*, 2002.

Chapter 2 – Take Your Partners

M. C. Baker, T. K. Bjerke, H. U. Lampe & Y. O. Espmark. Sexual response of female yellowhammers to differences in regional song dialects and repertoire sizes. *Animal Behaviour*, 1987.

C. Bartsch, M. Weiss & S. Kipper. Multiple song features are related to paternal effort in common nightingales. *BMC Evolutionary Biology*, 2015.

C. Biard, N. Saulnier, M. Gaillard & J. Moreau. Carotenoid-based bill colour is an integrative signal of multiple parasite infection in blackbird. *Naturwissenschaften*, 2010.

K. L. Buchanan & C. K. Catchpole. Song as an indicator of male parental effort in the sedge warbler. *Proceedings of the Royal Society*, 2000.

K. L. Buchanan, C. K. Catchpole, J. W. Lewis & A. Lodge. Song as an indicator of parasitism in the sedge warbler. *Animal Behaviour*, 1999.

S. Dale, T. Amundsen, J. T. Lifjeld & T. Slagsvold. Mate sampling behaviour of female pied flycatchers: evidence for active mate choice. *Behavioral Ecology and Sociobiology*, 1990.

M. Griggio & H. Hoi. Only females in poor condition display a clear preference and prefer males with an average badge. *BMC Evolutionary Biology*, 2010.

L. J. Henderson, B. J. Heidinger, N. P. Evans & K. E. Arnold. Ultraviolet crown coloration in female blue tits predicts reproductive success and baseline corticosterone. *Behavioral Ecology*, 2013.

M. Hoi-Leitner, H. Nechtelberger & H. Hoi. Song-rate as a signal for nest-site quality in blackcaps *Sylvia atricapilla*. *Behavioral Ecology and Sociobiology*, 1995.

A. P. Møller. Female choice selects for male tail ornaments in the monogamous swallow. *Nature*, 1988.

A. P. Møller. Badge size in the house sparrow *Passer domesticus*. Effects on intra- and intersexual selection. *Behavioral Ecology and Sociobiology,* 1988.

T. Radesäter, S. Jakobsson, N. Andbjer & A. Bylin. Song rate and pair formation in the willow warbler, *Phylloscopus trochilus*. *Animal Behaviour*, 1987.

J. C. Senar & J. Figuerola. Brighter yellow blue tits make better parents.
Proceedings of the Royal Society, 2002.

T. Slagsvold & J. T. Lifjeld. Variation in plumage colour of the great tit *Parus major* in relation to habitat, season and food. *Journal of Zoology*, 1985.

J. Sundberg. Female yellowhammers (*Emberiza citrinella*) prefer yellower males: a laboratory experiment. *Behavioral Ecology and Sociobiology*, 1995.

Fewer

J. Sundberg. Parasites, plumage coloration and reproductive success in the yellowhammer, *Emberiza citrinella*. *Oikos*, 1995.

Chapter 3 – The Breeding Circle

R. V. Alatalo & A. Lundberg. Polyterritorial polygyny in the pied flycatcher *Ficedula hypoleuca* – evidence for the deception hypothesis. *Annales Zoologici Fennici*, 1984.

C. J. Bibby. Polygyny and breeding ecology of the Cetti's warbler *Cettia cetti*. *Ibis*, 1982.

T. R. Birkhead, F. M. Hunter & J. E. Pellatt. Sperm competition in the zebra finch, *Taeniopygia guttata*. *Animal Behaviour*, 1989.

Davies, N. B. *Dunnock Behaviour and Social Evolution*. Oxford: OUP. 1992.

A. Dixon, D. Ross, S. L. C. O'Malley & T. Burke. Paternal investment inversely related to degree of extra-pair paternity in the reed bunting. *Nature*, 1994.

S. C. Griffith, I. P. F. Owens & K. A. Thuman. Extra-pair paternity in birds: a review of interspecific variation and adaptive function. *Molecular Ecology*, 2002.

I. G. Henderson, P. J. B. Hart & T. Burke. Strict monogamy in a semi-colonial passerine: the jackdaw *Corvus monedula*. *Journal of Avian Biology*, 2000.

T. Lubjuhn, T. Gerken, J. Brün & J. T. Epplen. High frequency of extra-pair paternity in the coal tit. *Journal of Avian Biology*, 1999.

Turner, A. K. *The Swallow*. London: Hamlyn. 1994.

University of Bonn. Age increases chance of success as two-timer for coal tit males. *ScienceDaily*, 2007.

Fewer

J. H. Wetton & D. T. Parkin. An association between fertility and cuckoldry in the house sparrow, *Passer domesticus*. *Proceedings of the Royal Society*, 1991.

Chapter 4 – Competitive Exclusion

K. E. Arnold & R. Griffiths. Sex-specific hatching order, growth rates and fledging success in jackdaws *Corvus monedula*. *Journal of Avian Biology*, 2003.

Birkhead, T. R. *The Magpies*. London: T. & A. D. Poyser. 1991.

C. W. Clark & J. Ekman. Dominant and subordinate fattening strategies: a dynamic game. *Oikos*, 1995.

W. Cresswell. Interference competition at low competitor densities in blackbirds *Turdus merula*. *Journal of Animal Ecology*, 1997.

D. W. Gibbons. Hatching asynchrony reduces parental investment in the jackdaw. *The Journal of Animal Ecology*, 1987.

M. Hake. Fattening strategies in dominance-structured greenfinch (*Carduelis chloris*) flocks in winter. *Behavioral Ecology and Sociobiology*, 1996.

I. Krams. Mass-dependent take-off ability in wintering great tits (*Parus major*): comparison of top-ranked adult males and subordinate juvenile females. *Behavioral Ecology and Sociobiology*, 2002.

R. K. Murton, A. J. Isaacson & N. J. Westwood. The significance of gregarious feeding behaviour and adrenal stress in a population of wood-pigeons *Columba palumbus*. *Journal of Zoology*, 1971.

Newton, I. *Finches*. London: Collins New Naturalist. 1972.

M. Orell & E. J. Belda. Delayed cost of reproduction and senescence in the willow tit *Parus montanus*. *Journal of Animal Ecology*, 2002.

J. L. Quinn & W. Cresswell. Personality, anti-predation behaviour and behavioural plasticity in the chaffinch *Fringilla coelebs*. *Behaviour*, 2005.

L. Ratcliffe, D. J. Mennill & K. A. Schubert. Social dominance and fitness in black-capped chickadees. In Otter, K. A. *Ecology and Behaviour of Chickadees and Titmice*. Oxford: OUP. 2007.

J. C. Senar. Allofeeding in Eurasian siskins (*Carduelis spinus*). *Condor*, 1984.

University of Exeter. Feed the birds: winter feeding makes for better breeding. *ScienceDaily*, 2008.

R. Zamora, J. A. Hodar & J. M. Gómez. Dartford warblers follow stonechats while foraging. *Ornis Scandinavica*, 1992.

Chapter 5 – Death and Declines

C. P. Bell, S. W. Baker, N. G. Parkes & M. L. Brooke. The role of the Eurasian sparrowhawk (*Accipiter nisus*) in the decline of the house sparrow (*Passer domesticus*) in Britain. *The Auk*, 2010.

Brown, L. *British Birds of Prey*. London: Collins New Naturalist. 1976.
S. Gooch, S. R. Baillie & T. R. Birkhead. Magpie *Pica pica* and songbird populations. Retrospective investigation of trends in population density and breeding success. *Journal of Applied Ecology*, 1991.
D.W. Groom. Magpie *Pica pica* predation on blackbird *Turdus merula* nests in urban areas. *Bird Study*, 1993.
S. E. Newson *et al.* Population change of avian predators and grey squirrels in England: is there evidence for an impact on avian prey populations? *Journal of Applied Ecology*, 2010.
R. L. Thomas, M. D. E. Fellowes & P. J. Baker. Spatio-temporal variation in predation by urban domestic cats (*Felis catus*) and the acceptability of possible management actions in the UK. *PLoS One*, 2012.
D. L. Thomson & R. E. Green. The widespread declines of songbirds in rural Britain do not correlate with the spread of their avian predators. *Proceedings of the Royal Society*, 1998.
M. Woods, R. A. McDonald & S. Harris. Predation of wildlife by domestic cats *Felis catus* in Great Britain. *Mammal Review*, 2003.

Chapter 6 – Repose

N. W. F. Bode, D. W. Franks & A. J. Wood. Limited interactions in flocks: relating model simulations to empirical data. *Journal of the Royal Society Interface*, 2010.
Feare, C. *The Starling*. Oxford: OUP. 1984.
A. McGowan, S. P. Sharp, M. Simeoni & B. J. Hatchwell. Competing for position in the communal roosts of long-tailed tits. *Animal Behaviour*, 2006.
D. J. G. Pearce, A. M. Miller, G. Rowlands & M. S. Turner. Role of projection in the control of bird flocks. *Proceedings of the National Academy of Sciences*, 2014.
I. R. Swingland. The social and spatial organization of winter communal roosting in rooks (*Corvus frugilegus*). *Journal of Zoology*, 1977.

Chapter 7 – Secrets and Robins

M. Cuadrado. Why are migrant robins (*Erithacus rubecula*) territorial in winter? The importance of the anti-predatory behaviour. *Ethology Ecology & Evolution*, 1997.
R. A. Fuller, P. H. Warren & K. J. Gaston. Daytime noise predicts nocturnal singing in urban robins. *Biology letters*, 2007.
Harper, D. G. C. The vanishing robin mystery. *Bird Watching magazine*, 2003.
A. Hoelzel. Song characteristics and response to playback of male and female robins *Erithacus rubecula*. *Ibis*, 1986.
Lack, D. *The Life of the Robin*. London: Witherby. 1943.
Lack, D., Harper, D. and Lack, P. *The Life of the Robin*. London: Pallas Athene. 2016.
W. Wiltschko & R. Wiltschko. Magnetic compass of European robins. *Science*, 1972.

Chapter 8 – On the Move

F. Bairlein *et al.* Cross-hemisphere migration of a 25g songbird. *Royal Society Publishing*, 2012.
C. J. Bibby & R. E. Green. Autumn migration strategies of reed and sedge warblers. *Ornis Scandinavica*, 1981.
Glutz von Blotzheim, U.N. & Bauer, K.M. *Handbuch der Vögel Mitteleuropas, Volume 9*. Weisbaden: Akademische Verlagsgesellschaft. 1980.
K. J. Park, M. Rosén & A. Hedenström. Flight kinematics of the barn swallow (*Hirundo rustica*) over a wide range of speeds in a wind tunnel. *Journal of Experimental Biology*, 2001.
www.bto.org/science/migration/tracking-studies/cuckoo-tracking

Chapter 9 – Finding the Way

P. Berthold & U. Querner. Genetic basis of migratory behavior in European warblers. *Science*, 1981.

S. T. Emlen. Migratory orientation in the indigo bunting, *Passerina cyanea*. Part II: Mechanism of celestial orientation. *The Auk*, 1967.

E. Gwinner & W. Wiltschko. Endogenously controlled changes in migratory direction of the garden warbler, *Sylvia borin*. *Journal of Comparative Physiology*, 1978.

G. Kramer. Experiments on bird orientation. *Ibis*, 1952.

A. Möller, S. Sagasser, W. Wiltschko & B. Schierwater. Retinal cryptochrome in a migratory passerine bird: a possible transducer for the avian magnetic compass. *Naturwissenschaften*, 2004.

R. Muheim, J. B. Phillips & S. Åkesson. Polarized light cues underlie compass calibration in migratory songbirds. *Science*, 2006.

Newton, I. *Migration Ecology of Birds*. London: Elsevier/Academic Press. 2008.

E. G. F. Sauer. Further studies on the stellar orientation of nocturnally migrating birds. *Psychological Research*, 1961.

Chapter 10 – Recreating the Pastoral Symphony

M. R. Ballintijn & C. Ten Cate. Variation in number of elements in the perch-coo vocalization of the collared dove (*Streptopelia decaocto*) and what it may tell about the sender. *Behaviour*, 1999.

T. K. Bjerke & T. H. Bjerke. Song dialects in the redwing *Turdus iliacus*. *Ornis Scandinavica*, 1981.

Cell Press. Cities change the songs of birds. *ScienceDaily*, 2006.

Davies, N. *Cuckoo, Cheating by Nature*. London: Bloomsbury. 2015.

J. C. Guyomarc'h *et al. Coturnix coturnix*. Quail. *BWP Update*, 1998.

W. Halfwerk *et al.* Low-frequency songs lose their potency in noisy urban conditions. *Proceedings of the National Academy of Sciences*, 2011.

M. L. Hunter & J. R. Krebs. Geographical variation in the song of the great tit (*Parus major*) in relation to ecological factors. *Journal of Animal Ecology*, 1979.

H Löhrl. Untersuchungen am kuckuck, *Cuculus canorus* (Biologie, Ethologie und Morphologie). *Journal für Ornithologie*, 1979.

M. P. Lombardo, H. W. Power & P. C. Stouffer. Egg removal and intraspecific brood parasitism in the European starling (*Sturnus vulgaris*). *Behavioral Ecology and Sociobiology*, 1989.

A. Lynch & A. J. Baker. A population memetics approach to cultural evolution in chaffinch song: differentiation among populations. *Evolution*, 1994.

R. Pfenning *et al.* Convergent transcriptional specializations in the brains of humans and song-learning birds. *Science*, 2014.

T. Roth, P. Sprau & R. Schmidt. Sex-specific timing of mate searching and territory prospecting in the nightingale: nocturnal life of females. *Proceedings of the Royal Society*, 2009.

W. H. Thorpe. The learning of song patterns by birds, with especial reference to the song of the chaffinch *Fringilla coelebs*. *Ibis*, 1958.

H. Williams, I. I. Levin, D. R. Norris & A. E. M. Newman. Three decades of cultural evolution in Savannah sparrow songs. *Animal Behaviour*, 2013.

Chapter 11 – A Thousand Cuts

Wood warbler declines in New Forest. Tony Davis, *pers comm.*

Index